Global Warming and the Asian Pacific

ACADEMIA STUDIES IN ASIAN ECONOMIES

Series Editor: Sheng-Cheng Hu, *Director, Institute of Economics, Academia Sinica, Taiwan*

Titles in the series include:

Economic Efficiency and Productivity Growth in the Asia-Pacific Region
Edited by Tsu-Tan Fu, Cliff J. Huang and C.A. Knox Lovell

Food Security in Asia
Economics and Policies
Edited by Wen S. Chern, Colin A. Carter and Shun-Yi Shei

The Economic Analysis of Substance Use and Abuse
The Experience of Developed Countries and Lessons for Developing Countries
Edited by Michael Grossman and Chee-Ruey Hsieh

Productivity and Economic Performance in the Asia-Pacific Region
Edited by Tsu-Tan Fu, Cliff J. Huang and C.A. Knox Lovell

The Economics of Health Care in Asia-Pacific Countries
Edited by Teh-Wei Hu and Chee-Ruey Hsieh

Monetary Policy and Taiwan's Economy
Edited by Gerald P. Dwyer, Jr., Jin-Lung Lin, Jia-Dong Shea and Chung-Shu Wu

Global Warming and the Asian Pacific
Edited by Ching-Cheng Chang, Robert Mendelsohn and Daigee Shaw

Global Warming and the Asian Pacific

Edited by

Ching-Cheng Chang

Research Fellow, Institute of Economics, Academia Sinica, Taipei, Taiwan and Professor, Department of Agricultural Economics, National Taiwan University, Taiwan, ROC

Robert Mendelsohn

Edwin Weyerhaeuser Davis Professor, School of Forestry and Environmental Studies, Yale University, New Haven, USA

Daigee Shaw

Research Fellow, Institute of Economics, Academia Sinica, Taipei, Taiwan, ROC

ACADEMIA STUDIES IN ASIAN ECONOMIES

Edward Elgar

Cheltenham, UK • Northampton, MA, USA

Published by
Edward Elgar Publishing Limited
Glensanda House
Montpellier Parade
Cheltenham
Glos GL50 1UA
UK

Edward Elgar Publishing, Inc.
136 West Street
Suite 202
Northampton
Massachusetts 01060
USA

A catalogue record for this book
is available from the British Library

Library of Congress Cataloguing in Publication Data
Global warming and the Asian Pacific / edited by Ching-Cheng Chang, Robert Mendelsohn, Daigee Shaw.
p. cm. — (Academia studies in Asian economies)
Includes bibliographical references and index.
1. Global warming. 2. Greenhouse effect, Atmospheric—East Asia. 3. Carbon monoxide—Environmental aspects—East Asia. 4. Climatic changes—Economic aspects—East Asia. I. Chang, Ching-Cheng, 1957– II. Mendelsohn, Robert O., 1952– III. Shaw, Daigee, 1951– IV. Series.

QC981.8.G56G575 2003
363.738'74'095—dc21

2003051612

ISBN 1 84376 419 9

Printed and bound in Great Britain by MPG Books Ltd, Bodmin, Cornwall

Contents

Tables

Figures

Contributors

Ching-Cheng Chang, Research Fellow, Institute of Economics, Academia Sinica and Professor, Department of Agricultural Economics, National Taiwan University, Taipei, Taiwan, ROC.

Chi-Chung Chen, Assistant Professor, Department of Applied Economics, National Chun-Hsin University, Taichung, Taiwan, ROC.

Carolyn Fischer, Fellow, Resources for the Future, Washington, DC, USA.

Frédéric Ghersi, Researcher and Scholar, Centre International de Recherche sur l'Environnement et le Développement (CIRED), Paris, France.

Shih-Hsun Hsu, Professor, Department of Agricultural Economics, National Taiwan University, Taipei, Taiwan, ROC.

Chung-Huang Huang, Professor, Department of Economics, National Tsing Hua University, Hsin Chu, Taiwan, ROC.

Ekko C. van Ierland, Professor, Environmental Economics and Natural Resources Group, Wageningen University, The Netherlands.

Kozo Ishida, Professor, Faculty of Economics, Rissho University, Tokyo, Japan.

Dale W. Jorgenson, Frederic E. Abbe Professor of Economics, Department of Economics, Harvard University, Boston, MA, USA.

Raymond Kopp, Senior Fellow, Energy and Natural Resources Division, Resources for the Future, Washington, DC, USA.

Joram Krozer, Research Director Cartesius Institute, University of Twente, CSTM/Cartesius Institute, The Netherlands.

Masahiro Kuroda, Professor, Faculty of Business and Commerce, Keio University and Vice President, Keio University, Tokyo, Japan.

Rianne de Leeuw, Researcher, University of Twente, CSTM/Cartesius Institute, The Netherlands.

Ping-Cheng Li, Professor, Department of Political Economy, National Sun Yat-Sen University, Kaohsiung, Taiwan, ROC.

Chi-Yuan Liang, Research Fellow, Institute of Economics, Academia Sinica, Taipei, Taiwan, ROC.

Hsing-Hua Lin, Assistant Professor, Department of Finance, Ching-Yun Institute Technology, Jung-Li, Taiwan, ROC.

Gang Ma, Ph.D. student, School of International Trade and Finance, Brandeis University, Waltham, Massachusetts, USA.

Bruce A. McCarl, Regents Professor, Department of Agricultural Economics, Texas A&M University, College Station, Texas, USA.

Warwick J. McKibbin, Professor, Australian National University, Australia and Senior Fellow, The Brookings Institution, USA.

Robert Mendelsohn, The Edwin Weyerhaeuser Davis Professor, Yale School of Forestry and Environmental Studies, Yale University, New Haven, CT, USA.

Richard Morgenstern, Senior Fellow, Quality of the Environment Division, Resources for the Future, Washington, DC, USA.

Koji Nomura, Associate Professor, Keio Economic Observatory, Keio University, Tokyo, Japan.

William Pizer, Fellow, Quality of the Environment Division, Resources for the Future, Washington, DC, USA.

Vu Xuan Quang, Guest Researcher, Department of Economics, Vienna University of Economics and Business Administration, Vienna, Austria.

Hikaru Sakuramoto, Professor, Faculty of Business and Commerce, Keio University, Tokyo, Japan.

Uwe A. Schneider, Assistant Professor, Research Unit Sustainability and Global Change, Center for Marine and Climate Research, Departments of Geosciences and Economics, Hamburg University, Hamburg, Germany.

Daigee Shaw, Research Fellow, Institute of Economics, Academia Sinica, Taipei, Taiwan, ROC.

Masahiko Shimizu, Professor, Faculty of Economics, Keio University, Tokyo, Japan.

Naoko Takenaka, Researcher, Keio University, Tokyo, Japan.

Ayu Washizu, Professor, School of Social Sciences, Waseda University, Tokyo, Japan.

Yuxin Zheng, Senior Research Fellow and Deputy Director of Quantitative and Technical Economics, Chinese Academy of Social Sciences, Beijing, PR of China.

Foreword

This book emanated from a conference held in Taipei in December 2000. The conference was sponsored by the Taiwan Cement Company and held at the Institute of Economics at the Academia Sinica. Scholars from around the Pacific Rim were invited to talk about global warming and pollution control in the Asian Pacific. From this broader conference, selected papers were gathered that dealt with one of four topics concerning global warming: baseline emissions in the Asian Pacific, the cost of abating greenhouse gas emissions, the impacts of global warming on the Asian Pacific, and global warming policy. Although a great deal has been written about greenhouse gases in general, this is the first book that deals with this problem specifically in the Asian Pacific. The book contains some of the most detailed and up to date information available about global warming and the Asian Pacific.

As a collection of papers, the book has many authors from many countries. Many of the Asian Pacific countries are represented including China, Japan, Vietnam and Australia. Global warming experts from the United States and Europe have also made contributions. The book is clearly a multinational effort and reflects just how much interest there is in global warming from around the world.

Each paper in this book went through a rigorous review process. The papers were carefully reviewed by the editors who made several comments on each paper, and the authors responded with many improvements. The resulting collection of papers is consequently an edited volume.

New information concerning global warming is accruing at a rapid rate making it increasingly difficult to stay on top of all advances. As we learn more about the origin of greenhouse gases, their control and their impacts, our ability to design new and more effective policies will certainly increase. One point that this book makes very clear however is that every country in the world will be affected. Every country contributes to greenhouse gas emissions, every country has the potential to reduce their emissions, and every country will be affected if the climate of the world changes. Each region of the world must come to terms with what global warming entails for them from an abatement perspective, from an impact perspective, and from a policy perspective. This book contributes to our understanding by shedding light on the Asian Pacific. Not only should this be of great interest to this region, but the book also serves as a model for how regional perspectives could be captured across the world.

Acknowledgements

Chapter 7 was originally published in 'China CGE Model and Policy Analysis', Social Science Literature Publishing House, Beijing, 1999. Reprinted here by kind permission of Social Science Literature Publishing House.

1. Global warming and the Asian Pacific

Ching-Cheng Chang, Daigee Shaw and Robert Mendelsohn

1 INTRODUCTION

A great deal has been learned about greenhouse gases and climate change over the last two decades (Houghton et al., 1996; Watson et al., 1996; Bruce et al., 1996; Houghton et al., 2001; Metz et al., 2001; McCarthy et al., 2001). The research has revealed that greenhouse gases are a long-term problem that began over a century ago and will extend for centuries into the future. The research has also revealed that it is truly a global problem in that the emissions are in fact emanating from every country in the world and that the impacts will be felt in every corner of the world. Because of these two factors, greenhouse gas science and policy is clearly not a simple matter but is instead incredibly complex and difficult to fathom. People must think beyond narrow borders of space and beyond near-term actions and consequences. Studying climate change requires understanding across vast spaces and over very long time periods. It also requires understanding multiple disciplines in both the natural and social sciences. It is no wonder that few, if any, individuals have a grasp of the entire phenomenon.

We begin in this chapter with a brief overview of the literature on the global effects of greenhouse gases and their control (see Houghton et al., 2001). Greenhouse gases are emitted as a byproduct of burning fossil fuels and to a lesser degree from agricultural and industrial processes. The gases include carbon dioxide, nitrogen oxides, methane and numerous halocarbons. Every country in the world contributes to greenhouse gas emissions. Because the gases mix in the atmosphere, the source of the emissions does not matter. All that matters is the global quantity of emissions. Every country needs to be aware of how they are contributing to this global total. This book provides projections of future greenhouse gas emissions from the Asian Pacific.

Greenhouse gas emissions have been rising since the beginning of the industrial revolution and are expected to continue to rise in the absence of any control policies as long as the world continues to burn fossil fuels. The emissions could continue for centuries to come. These emissions have already led to rising concentrations of these gases and to subtle temperature shifts across the planet. However, the concern about greenhouse gases lies not in what has already happened but rather in what might happen in the future as temperatures potentially rise dramatically.

Greenhouse gas emissions can be controlled. Economists believe there is a range of options for controlling these gases that increase in cost the more one reduces emissions. Some analysts argue that in addition, some reductions are free – they would have no cost. The global cost of controlling emissions is hotly debated. However, it is clear that the cost of abatement will be different for individual countries depending upon their fossil fuel reserves, their energy efficiency and their future development. How much each country must spend to control emissions is clearly of great national interest. A second topic covered in this book is the abatement costs for countries in the Asian Pacific.

Because greenhouse gases are expected to alter the climate of the entire earth, every country will be impacted. However, the impacts are not likely to be the same for each country. First, the climate is not expected to change uniformly. Some countries might experience a greater or lesser change in temperature and precipitation might change dramatically from country to country. Second, each country begins with a different climate. Warming will fall more heavily on countries that are already too hot. The impact of changes in precipitation will depend greatly on how much precipitation is currently occurring. Third, only certain sectors will be affected by climate change. The magnitude of impacts will depend on the relative size of these climate sensitive sectors in each country. A third topic raised in this book concerns how the Asian Pacific countries will be affected by different climate scenarios.

A final topic raised in the book concerns global warming policies. Because this problem requires coordination across all countries in the world, it raises new institutional questions. How can pollution be controlled in each country both efficiently and equitably? Although these policies are not specific to Asian Pacific countries, they clearly will have implications for each country in the region. The policies will determine abatement costs and future damages. Every country must become aware of what is at stake and what is in their best interest in this international negotiation.

This book contributes to our understanding about climate change by examining the Asian Pacific region in detail. The Asian Pacific is the home of many of the contributors. However, we have selected the Asian Pacific as the focus of this book because it has received little attention in the past so that the research in this book sheds many new and interesting insights. Further, by examining many important issues from a regional perspective, the book provides a useful blueprint for how regional studies could be conducted elsewhere.

2 REVIEW OF GLOBAL ISSUES

For the last century, scientists have speculated whether greenhouse gas emissions might cause global warming. As the science has evolved, global warming has progressed from a strictly scientific inquiry into a global policy concern. Some of the links between man's activities and the impacts of climate change are well understood and some remain clouded in uncertainty.

It is well understood that the combustion of fossil fuels for energy causes the emission of vast quantities of carbon dioxide (Watson et al., 1996; Metz et al., 2001). The global growth in agricultural land also contributes to this carbon dioxide account by converting natural forests with extensive carbon into farming land with much lower carbon stocks (Watson et al., 1996; Metz et al., 2001). Industrial and agricultural practices are sources of additional greenhouse gases such as methane and nitrogen oxides (Watson et al., 1996; Metz et al., 2001). It is not well understood how these emissions might change over time. The emissions are expected to grow with the growth of the world's population and the expansion of the world economy. Although we are able to project how these forces are likely to change in the near term, projections several decades and especially centuries into the future cannot be done with precision. Economic growth might change for unforeseeable reasons. Another source of uncertainty concerns the role future technological changes might have. Improvements in energy efficiency have been proceeding steadily over history but the continued pace of these innovations into the next century is difficult to determine.

It is well known that greenhouse gas emissions are causing the concentrations of greenhouse gases in the atmosphere to rise (Houghton et al., 1996, 2001). Records of greenhouse gas concentrations suggest that they are increasing by approximately one half the amount of annual emissions. It is clear that some of the emissions are being reabsorbed into the marine and terrestrial environment. The exact role of every sink is not yet understood. Nor is it clear how much of future emissions these sinks will continue to be able to absorb.

There is no question that greenhouse gases have the potential to increase future temperatures (Houghton et al., 1996, 2001). These gases permit visible light to pass through but they block infrared heat from escaping. The gases behave just as glass does in a greenhouse. One can look to our nearest neighbors in the solar system and observe that Venus is much hotter than

Earth because it has a much higher concentration of these gases but Mars is much colder because it has very little atmosphere.

What is not understood is exactly how Earth's climate will change with higher concentrations of greenhouse gases (Houghton et al., 1996, 2001). Many things will change with the direct increase of temperature from greenhouse gases. For example, the hydrological cycle will increase and that is expected to increase the amount of water in the atmosphere. Water is itself a greenhouse gas but the precise impact of this feedback depends on how the clouds change. Other subtle changes that will occur entail changes in ice and the biosphere that will change how much light is reflected back into space. At best, science can only predict a range of temperature and precipitation changes that might occur.

The timing of these climate changes is also uncertain (Houghton et al., 1996, 2001). The climate changes will clearly become more dramatic as greenhouse gas concentrations rise, but what we will actually experience each decade is not known with precision. There will be substantial delays between emissions and climate change. The primary lag in the system is the ocean. Although air warms relatively quickly, it is expected that it will take decades for the increased heat to warm the ocean. As the ocean warms, it will carry this heat further poleward and bring the earth to a higher equilibrium temperature.

The long lag between emissions and consequences has important implications for policy (Bruce et al., 1996). Policy makers must anticipate what emissions will do in the future. They will react too slowly if they wait to see what actually happens before acting. Policy makers must not be complacent just because the climate has not already become troublesome. They must decide whether the future changes are worth avoiding, not the past.

Unfortunately, it is difficult to forecast far into the future. Climate policies must consequently be decided despite large uncertainties about the future costs and especially the future benefits of control. One important link that is highly uncertain is the future distribution of climate changes across the planet. Scientists expect that the polar regions will experience the greatest amount of warming and the tropical regions the least. But the exact amount of warming that takes place in each location is not known as different climate models predict different patterns (Houghton et al., 1996, 2001). The geographic distribution of precipitation changes is even less clear. Although all climate models predict an overall increase in precipitation, how rainfall will change across the planet is very difficult to predict. Each model seems to predict different patterns of increases and decreases.

Even if the exact nature of future climate change were known, there remains considerable uncertainty surrounding what impacts these changes will cause (Watson et al., 1996; Bruce et al., 1996; McCarthy et al., 2001).

Analysts have identified parts of the economy that are more vulnerable: agriculture, forestry, energy, coastal resources and water. However, the impact of warming is likely to be different depending upon the size of these sectors in the future, the current climate in an area and the level of development. For example, if agriculture expands, farming impacts will be larger. If one were in a relatively cool region, warming might actually help agriculture whereas if one started in a relatively hot location, warming might be strictly harmful. If one had access to many technological substitutes for climate such as irrigation, mechanized agriculture and modern varieties, the effect of climate might be mild. On the other hand, if agriculture was more labor intensive and there were few choices for farmers to adapt, the impact of warming might be more severe. There are many factors that will affect the magnitude and even the direction of climate impacts on the economy.

In addition to affecting the economy, global warming is also expected to affect the quality of life. These quality of life effects, however, are even more uncertain than the economic impacts. First, it is not completely clear how climate change will change the quality of life (Watson et al., 1996; McCarthy et al., 2001). Second, we do not have good measures of what values should be placed on these quality of life changes (Bruce et al., 1996). For example, there is little doubt that ecosystems will be affected. Ecosystem models predict widespread shifts in biomes across the planet as well as changes in productivity. These shifts are likely to affect local and global landscapes, recreation and endangered species. It is not clear, however, what weight to give these changes. That is, it is not clear whether these changes are relatively minor or extremely important. In addition, there are other effects. Ecosystem changes may have health implications as vector-borne diseases find new territories to invade. Health might be directly affected by changes in heat-related incidents, shifts in seasonal patterns and changes in pollution. People will also be clearly affected by the changes in the weather they experience.

The global literature suggests powerful economic forces will be at work with respect to climate policy. The literature implies that abating greenhouse gases is possible but it will be expensive (Bruce et al., 1996; Metz et al., 2001). Mild programs may only cost hundreds of millions of dollars per year but they will only slow warming slightly. More active programs that substantially reduce warming will cost a lot more, from 2 to 5 trillion dollars, depending on their design.

Economic assessments of the benefits of controlling greenhouse gases are muted (Bruce et al., 1996). By avoiding potential damages, there is no question that benefits exist. The problem is that the benefits appear to be relatively small for aggressive mitigation programs. Aggressive programs involve large mitigation costs that begin immediately but the benefits may be delayed by several decades or even a century. The present value of the benefits are small compared to the present value of the costs. The economic assessments of climate change thus suggest modest low-cost control programs that only slow warming slightly.

The distribution of costs and benefits is also important. Abatement costs will not be spread evenly across countries (Bruce et al., 1996; Metz et al., 2001). Nations that have already encouraged energy efficiency will find it difficult to carry out further reductions. Countries that use substantial coal will find they must make the largest changes. Developing countries potentially face the largest burdens as they wish to expand their economies dramatically and yet are the least energy efficient. Depending upon how policies are designed, greenhouse gas abatement could be very expensive for selected countries.

The benefits of controlling greenhouse gases are also not distributed uniformly (Mendelsohn et al., 2000). Polar countries may benefit if the earth warms. Tropical countries are expected to be the most vulnerable to climate change. Most of the developed world stands little to gain from controlling climate change. Most of the benefits will fall across developing countries. The distribution of impacts makes negotiating a global treaty very difficult. The countries responsible for most of the emissions have the least to gain from control. The countries that will gain the most from control have not caused the bulk of emissions nor are they in an economic position to pay for them.

The politics of constructing a meaningful global agreement to control greenhouse gases should not be underestimated. Each country has a different perspective concerning how it would like to control its own emissions and how it would like others to control theirs. Each country has a different perspective about impacts in their own country and impacts that fall on others. The problem is a very long-term one that requires very long-term solutions. There are few arrangements that could bind nations over the time horizons needed. Cheating and therefore enforcement of international agreements to control emissions is likely to be a serious problem. There is a lot of work to be done before effective agreements will be reached.

This book does not solve these difficult and intractable problems. The book, however, does address these issues and sheds new light on them. In particular, the book should make it clear to the Asian Pacific countries that they must become involved in climate change negotiations and preparations. They will not be exempt from this issue. Hopefully, other countries around the world will see this regional study and realize that they too must get involved. Although climate change will be a difficult problem to resolve over the next century, every country must get ready for global climate policies that will affect them and for climate changes that will occur whether they address emissions or not.

3 OVERVIEW OF THE BOOK

This book examines the problem of global warming from the perspective of the Asian Pacific countries. The unprecedented economic and population growth in the Asian Pacific over the past two decades has contributed significantly to the increases in energy use and greenhouse gas emissions. According to Bando (1996), Asia releases an estimated 8.4 billion metric tons of carbon dioxide each year from fossil fuel use, cement manufacturing and land use changes, while its emissions of methane represent half the global total. Consequently, the Asian Pacific region is becoming more important in the global response to climate change. The book is divided into four sections, each of which addresses a different aspect of the climate change problem: baseline emissions, mitigation costs, impact damages and policy solutions.

The first set of chapters deals with estimating baseline emissions. These studies predict uncontrolled greenhouse gas emissions over the near term, that is, what would happen to emissions if no control policies are adopted. Although this sounds like a simple exercise, it requires an in-depth understanding of energy use in each country and how energy use is changing. Four studies cover baseline emissions over the next decade. Quang begins this section with a relatively simple analysis of Vietnam's likely carbon emissions. Li, Hsu, Huang and Lin rely upon a much more complex and detailed economic–energy model to predict Taiwan's likely carbon emissions. Kuroda and Nomura use a dynamic Computable General Equilibrium (CGE) model to compute what Japan's emissions will be. Sakuramoto, Ishida, Shimizu, Washizu and Takenaka examine the trade links across each country in the region. Sakuramoto et al. demonstrate that the emissions in each country are affected by trade and that how they are counted depends upon how you assign emissions to intermediate and final products. Although the sophistication of the tools varies considerably across these studies, they all yield consistent predictions of growing emissions in every country.

The second set of chapters in this book estimate abatement costs. Abatement costs depend upon the growth of the economy and its ability to move away from emissions. Economies with more emissions such as China generally have more opportunities to abate, to remove emissions. Economies with more fuel choices, more substitutes for carbon-rich fuels like coal, will face lower costs. Economies with more opportunities to improve their energy efficiency will also face lower costs. The first study in this section by Liang and Jorgenson uses a complex engineering–economic model to predict the cost of different abatement strategies for Taiwan. Chapter 7, by Zheng and Ma, uses a static CGE model to predict abatement costs in mainland China. This is a particularly difficult task given the mixture of private and public institutions that vie for resources in China. Kuroda and Nomura use a dynamic CGE model to estimate the abatement costs for Japan. This is a sophisticated tool that allows the economy to find the least-cost way of achieving specified abatement levels over time. Finally, Schneider and McCarl explore whether biomass fuels from agriculture can help with abatement. All of these models confirm that abatement is possible but it will be expensive. The more abatement that is desired and the faster the abatement targets must be met, the higher the costs will be. The studies also suggest that the more energy efficient the economy (Japan), the higher the costs of further abatement. Countries such as mainland China, in contrast, with their low energy efficiency and vast use of coal have many low-cost opportunities to reduce carbon emissions.

The third set of studies in this book concerns the impacts of climate change in the Asian Pacific. Chang and Chen present a model of agriculture in Taiwan. Using information about individual crops, they explore how Taiwan's farming system will be affected by warming. Mendelsohn and Shaw present an empirical health study that examines how temperature affects the mortality rate of the elderly in the United States. The study shows that mortality rates go up as the monthly variance in temperature increases but not as the temperature level increases. Finally, Mendelsohn presents results from his global model of economic impacts for the Asian Pacific. This chapter predicts that the more northern countries in the region (Japan and China) are likely to benefit from warming. Countries that are in the tropical zone however will get too warm and will suffer damages. The biggest damages are from agricultural losses, especially in Australia, Indonesia and Thailand. Non-market damages may also be large across all the warmer nations in the region. The results indicate that the impacts of warming will not be uniform across the region as some countries may actually benefit while others are likely to be harmed.

The final set of studies in the book concern global warming policies. These policies apply to the region but generally concern the entire globe. Kopp, Morgenstern, Pizer and Ghersi recommend carbon targets for each country but to allow countries to break those targets if abatement costs get above a pre-specified level. Although the literature suggests that the optimal tax rate should be in the neighborhood of $10 per ton, the authors arbitrarily press for $50 per ton as a maximum price (Nordhaus and Boyer, 2000; Pearce et al., 1996). McKibbin introduces a similar strategy that involves a mixture of price and quantities. McKibbin allows countries to distribute additional permits whenever the price of carbon rises above a specified global price. Van Ierland, de Leeuw and Krozer examine joint implementation under the Kyoto agreement. They find that although the countries that have agreed to limit emissions could profitably purchase reductions from non-participating countries, it will be difficult to design an effective program. There will be many times when project level emission reductions may not reduce a country's emissions by very much. Fischer concentrates on designing efficient control policies for an individual country. She notes that price policies will tend to be more efficient than control policies and specifically presses for a tax on carbon emissions. This set of chapters begins to tackle the difficult problem of establishing effective policy tools to control warming. Although none of the authors provides a complete answer to all the problems facing global control of greenhouse gases, all the authors provide fresh new ideas to consider. It will take such a refreshing combination of originality and expertise to fashion a workable international program.

REFERENCES

Bando, A. (1996), 'Strategic Planning and Adoption of Greenhouse Gas Mitigation Options in Asia', *Fuel and Energy Abstracts*, **37**, pp. 308.

Bruce, J., H. Lee and E. Haites (eds) (1996), *Climate Change 1995: Economic and Social Dimensions of Climate Change*, Contribution of Working Group III to the Second Assessment Report of the Intergovernmetnal Panel on Climate Change, Cambridge: Cambridge University Press.

Houghton, J., L. Meira Filho, B. Callander, N. Harris, A. Kattenberg and K. Maskell (eds) (1996), *Climate Change 1995: The Science of Climate Change*, Second Assessment Report of the Intergovernmental Panel on Climate Change, Cambridge: Cambridge University Press.

Houghton, J., Y. Ding, D. Griggs, M. Nogeur, P. van der Linden, X. Dai, K. Maskell and C. Johnson (eds) (2001), *Climate Change 2001: The Scientific Basis*, Working Group I Third Assessment Report of the Intergovernmental Panel on Climate Change, Cambridge: Cambridge University Press.

McCarthy, J., O Canziani, N. Leary, D. Dokken and K. White (2001), 'Climate Change 2001: Impacts, Adaptation, and Vulnerability', Third Assessment Report of the Intergovernmental Panel on Climate Change, Cambridge: Cambridge University Press.

Mendelsohn, R., W. Morrison, M. Schlesinger and N. Adronova (2000), 'Country-Specific Market Impacts from Climate Change', *Climatic Change*, **45**, pp. 553–69.

Metz, B., O. Davidson, R. Swart and J. Pan (2001), 'Climate Change 2001: Mitigation', Third Assessment Report of the Intergovernmental Panel on Climate Change, Cambridge: Cambridge University Press.

Nordhaus, W.D. and J. Boyer (2000), *Warming the World: Economic Models of Global Warming*, Cambridge, MA: The MIT Press.

Pearce, D., W. Cline, A. Achanta, S. Fankhauser, R. Pachauri, R. Tol and P. Vellinga (1996), 'The Social Cost of Climate Change: Greenhouse Damage and the Benefits of Control', in J. Bruce, H. Lee and E. Haites (eds) (1996), *Climate*

Change 1995: Economic and Social Dimensions of Climate Change, Cambridge: Cambridge University Press.
Watson, R., M. Zinyowera, R. Moss and D. Dokken (eds) (1996), *Climate Change 1995: Impacts, Adaptations, and Mitigation of Climate Change: Scientific-Technical Analysis*, Second Assessment Report of the Intergovernmental Panel on Climate Change, Cambridge: Cambridge University Press.

PART I

Baseline Emissions

2. Forecasting carbon dioxide emissions in Vietnam

Vu Xuan Quang

1 INTRODUCTION[1]

Energy consumption is increasing in developing countries, essentially to meet their goals for economic and social development. However, satisfying this development has been associated with significant environmental impacts (pollution of land, air and water), potential long-term global climate change (through the emissions of greenhouse gases, primarily carbon dioxide and methane), and social concerns (World Bank, 1998). Studies have shown that CO_2 emissions are important because of their impact on climate change. Most CO_2 emissions are due to fossil energy consumption.

Global warming is considered to be an important environmental problem. Many attempts have been made by international organizations and countries to negotiate control of greenhouse gas emissions. The Kyoto Protocol is an important milestone towards promoting a reduction in greenhouse emissions.[2] Developing countries did not commit to specific reductions,[3] primarily for two reasons: first, their priorities are economic growth and poverty reduction, and secondly, industrialized countries consume far more energy and thus produce more greenhouse gases. However, they see the Kyoto agreement as a test of how serious the world is about climate change (IISD, 1998). Importantly, research shows that, after 2015, developing countries will produce more than 50 per cent of the world's greenhouse gas emissions. Developing countries will therefore be part of the global solution. Furthermore, global warming and climate change cannot be addressed without the involvement of developing countries (IISD, 1998; UNDP and WRI, 1999).

Like energy consumption in many developing countries, energy consumption has increased in Vietnam over the last two decades, primarily in order to meet the country's social and economic development. Since the country shifted from a centrally planned to a market economy in 1986, the Vietnam economy has grown rapidly, achieving an annual growth rate of 8 to 9 per cent in recent years. Energy consumption has also noticeably increased because of the country's economic reform process. This has facilitated economic growth and economic reforms in Vietnam, but it has also increased the emissions of carbon dioxide. Assessing and forecasting CO_2 emissions is therefore an essential and necessary regulatory response. The assessment of CO_2 emissions requires Vietnam to determine which factors affect emissions and how these factors might change over time.

This chapter adopts a simple model, PGESC, an abbreviation of population growth (*p*), economic performance (GDP per capita, *g*), energy intensity (*e*), energy structure (*s*), and carbon intensity (*c*), to illustrate the relationship between the economy, energy and carbon dioxide emissions in Vietnam.[4] CO_2 emissions are assessed in Vietnam from 1980 onwards and forecasted to 2010. This study considers economic growth, population, energy consumption and energy conversion when examining the changes in CO_2 emissions. It investigates the changes in CO_2 emissions based on population growth, economic intensity, energy intensity, fossil intensity and carbon intensity. It also explains the changes in CO_2 emissions over the short term. It informs decision makers in Vietnam of the magnitude and trajectory of emissions in the country. It is consequently an important first step towards designing a Vietnamese policy to address global warming.

This chapter consists of five sections. The next section introduces the model, PGESC, the affecting factors, and the emissions coefficients. The third section presents the empirical results and the findings relating to the changes in CO_2 emissions. The fourth section illustrates trends and forecasts for CO_2 emissions in Vietnam to the year 2010. The last section concludes.

2 THE MODEL

This chapter adopts an energy engineering approach to predict carbon dioxide emissions in Vietnam. The PGESC model illustrates the relationship between the economy, energy and the environment. It is based on the Ehrlich equation,[5]

$$I = P * A * T$$

where *I* is the adverse environmental impact, *P* is the population, *A* is the consumption per capita, and *T* is the amount of resources required relative to one unit of consumption (Ehrlich and Ehrlich, 1990; Zhang, 1999). The equation is used as a proxy for a determinant of the environmental impact.

In the model, CO_2 emissions resulting from energy consumption in sector *i* are estimated by multiplying energy consumed by a carbon emissions coefficient,

$$C_i = E_{i,j} \times \alpha_{i,j}$$

where

C: CO_2 emissions (tons of CO_2)

E: Energy consumption (toe)[6]

α: Emissions coefficient (tCO_2/toe)

The estimation uses sets of data collected from Vietnam's annual energy balance and a set of emissions coefficients. The energy unit used in this study is toe (tons of oil equivalent, where 10^6 toe = 41.86 PJ).[7] CO_2 emissions are based on metric tons of CO_2, calculated from the equation: $tCO_2 = tC \times (44/12)$.[8]

Carbon dioxide emissions depend on five main factors: economic performance, population, energy consumption, energy conversion and carbon intensity. Emissions depend on: gross domestic product (G), population (P), energy consumption (E), fossil energy consumption (S), and carbon (related to fossil energy consumption) (C),

$$C = f(G, P, E_0, E, C^0) \tag{2.1}$$

where

C : Carbon dioxide emissions (tCO_2),

G : Gross Domestic Product (US Dollars),

P : Population (head),

E_0: Primary energy consumption (toe),

E : Fossil energy consumption (toe), and

C^0: Carbon (tC).

Changes in CO_2 emissions over time also depend on the changes of these variables. Equation (2.1) can be expressed in terms of time, t, as follows:

$$C_t = f_t(G_t, P_t, E_{0t}, E_t, C_t^0) \tag{2.2}$$

Based on the 'Kaya-Identity',[9] a simple equation connecting the factors can be written as:

$$C = P * \frac{G}{P} * \frac{E_0}{G} * \frac{E}{E_0} * \frac{C^0}{E} * \frac{C}{C^0} \tag{2.3}$$

Substituting p, g, e, s, c, and c^0 for P, (G/P), (E_0/G), (E/E_0), (C^0/E), and (C/C^0), respectively, equation (2.3) can be rewritten as follows:

$$C = p * g * e * s * c * c^0 \tag{2.4}$$

In equation (2.4), factor p is *population,* factor g is GDP per capita or *economic intensity*, and factor e is primary energy consumption per one unit of GDP or *energy intensity*. Similarly, factor s is fossil energy consumption per one unit of primary energy consumed, indicating the *fossil ratio* (or *fossil intensity*) of energy consumption, and factor c is carbon contained in one unit of fossil energy consumed or *carbon intensity*. c^0 is carbon dioxide related to one unit of carbon, which is equal to (44/12). From an energy economics perspective, equation (2.4) maintains that CO_2 emissions are significantly related to economic intensity, population, energy intensity, fossil intensity and carbon intensity.[10]

Taking natural logs and differences over time from equation (2.4), the change in CO_2 emissions can be found to be:

$$\Delta C = \Delta p + \Delta g + \Delta e + \Delta s + \Delta c + \Delta c^0 \tag{2.5}$$

Since $c^0 = C/C^0 = 44/12$ is constant, $\Delta c^0 = \text{zero}$ and

$$\Delta C = \Delta p + \Delta g + \Delta e + \Delta s + \Delta c \tag{2.6}$$

Equation (2.6) maintains that the changes in CO_2 emissions (ΔC) depend on the change in population (Δp), the change in economic intensity (Δg), the change in energy intensity (Δe), the change in the fossil ratio (Δs) and the change in carbon intensity (Δc). The change in economic intensity is concerned with economic growth, the structure of the economy and economic development policies. The change in energy intensity depends on the energy system, the industrial structure and energy conservation policies.

Together with the two important parameters above, the changes in the fossil ratio (or fossil intensity) and carbon intensity are closely associated with energy efficiency, energy conversion and environmental policies. The changes in the population are determined by demographics and socio-economic development. These five parameters determine changes in CO_2 emissions.

As shown in Figure 2.1, the PGESC model requires data on energy balances, socio-economic development, emissions coefficients and scenarios. Data on energy include detailed information on the supply and consumption of energy products. We group this data into six consumer groups based on final energy consumption. Data concerned with socio-economic issues comprise data on GDP, the economic structure and population. Data shown in the scenarios are concerned with economic growth, population growth, forecasts of energy consumption, as well as energy development plans. To illustrate the model, we examine a baseline and five alternative scenarios: energy conservation, hydro and renewable energy, switching coal for gas, switching oil for gas, and mixing these strategies.

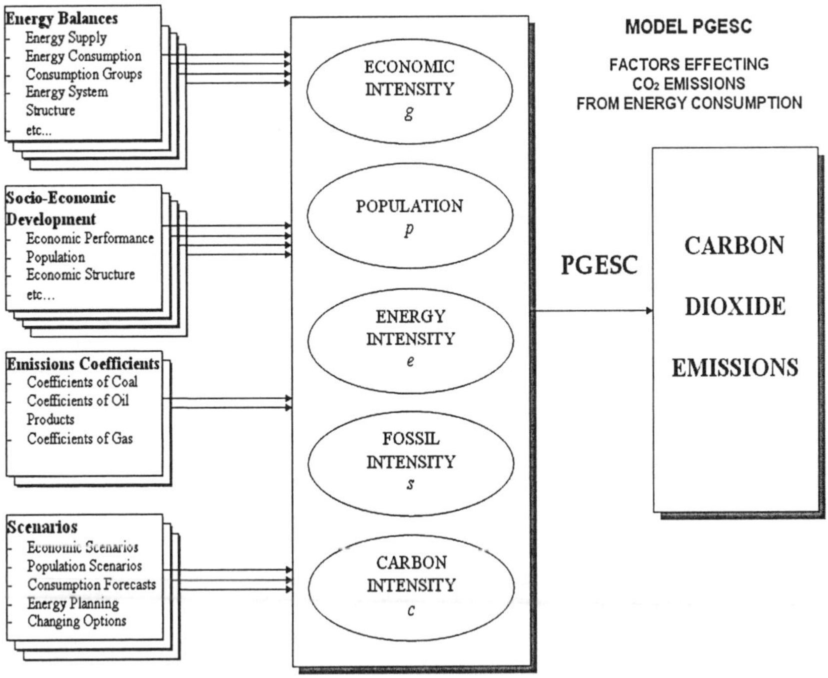

Figure 2.1 The structure of PGESC

To estimate carbon dioxide emissions from energy consumption, we need relevant emissions coefficients. Many organizations and research institutes[11] release the values of emissions coefficients. They are almost the same but may also be slightly different because of experimental conditions and the samples taken in labs. In this study, we use the emissions coefficients provided by the Intergovernmental Panel on Climate Change (IPCC), the Center on Energy and Environment of the UN Environmental Program (Riso Lab), and the Stockholm Environment Institute in Boston (SEI), since the values provided are most widely used in many countries to inventory carbon dioxide emissions.[12] Table 2.1 shows a detailed set of emissions coefficients converted to metric tons of CO_2 per ton of oil equivalent (tCO_2/toe).

Table 2.1 CO_2 emissions coefficients

Code	Energy	Coefficient (kg CO_2/GJ)			Coefficient (ton CO_2/toe)
cij		IPCC[1]	SEI[2]	Selected[3]	Converted[4]
		COAL			
c11	Coal Anthracite	98.3		98.3	4.1
c12	Coal Lignite		100.6	100.6	4.2
c13	Coke	108.2		108.2	4.5
c14	Other Coal[5]				4.2
		OIL			
c21	Crude Oil		73.1	73.1	3.1
c22	Gasoline	69.3	70.6	70.0	2.9
c23	Gasoline-TC1			70.0	2.9
c24	Fuel Oil	77.4	74.6	76.0	3.2
c25	Diesel Oil	74.1	74.6	74.4	3.1
c26	Kerosene		72.1	72.1	3.0
c27	LPG		66.0	66.0	2.8
c28	Other Oil	73.3		73.3	3.1
		GAS			
c31	Natural Gas	56.1	52.8	54.5	2.3
c32	Associated Gas				2.3

Notes and Sources:

1 IPCC and UNEP Center on Energy and Environment (1999).

2 SEI (Stockholm Environment Institute): A Guide to Environmental Analysis for Energy Planners (1995).

3 Coefficients are principally selected based on IPCC.

4 The conversion factor used is: 1 mtoe = 41868×GJ (General Conversion Factors for Energy, OECD).

5 Coefficient of other coal is assumed to be 4.2 (equal to Lignite).

3 EMPIRICAL RESULTS

3.1 Historical Data and Movements of Relative Factors

Vietnam has achieved a considerably high economic growth rate since the early 1990s. Whereas the country's per capita GDP has been as low as US$350 (World Bank, 2000),[13] the economy achieved an annual growth rate of more than 8 per cent during 1992–97. As a result of these economic driving forces, the country's energy consumption has increased rapidly, with an average growth rate of about 15 per cent. The population has grown at about 2.2 per cent per year, but it appears to have been slowing down slightly. Table 2.2 reviews the historical data for GDP, population, energy consumption and carbon dioxide emissions in Vietnam since 1980.

As shown in Table 2.2, primary energy at first grew slowly but more recently it has been growing at 15 per cent per year. Fossil energy use followed the same pattern except for a noticeable drop in 1989. Carbon dioxide emissions grew slowly in the 1980s but more rapidly in the 1990s. CO_2 emissions in 1980 were about 16 million tons, but by 1996, they had doubled.

The estimation of CO_2 emissions in Vietnam indicates a largely increasing contribution from oil in the 1990s. In the 1980s, oil and coal contributed 34 per cent and 66 per cent of the total emissions, respectively. By the mid 1990s, oil consumption accounted for nearly 60 per cent of the total emissions. In 1995, oil made up 57 per cent of the total emissions, while coal and gas had shares of 42 per cent and 1 per cent, respectively. The emissions from oil more than doubled in the late 1990s compared to 1990 and were more than three times higher than 1980 emissions.

Under the bottom-up estimation method, the CO_2 emissions changed structurally over the last decade, most noticeably due to the increasing participation of hydro-energy in power generation as well as the increase in energy consumption in transportation and households. During the second half of the 1980s, CO_2 emissions in power generation were around 30 per cent of the total emissions. However, the emissions decreased to below 20 per cent by the late 1990s because of the operation of the Hoa-Binh hydropower plant with its capacity of 2400 MW in 1989. This dam is the reason for the sudden sharp change in the energy patterns of Vietnam in 1989. The bottom-up structure in 1995 indicates that the emissions in industry and transportation were 42 per cent and 23 per cent, respectively, whereas the emissions from power generation were only 19 per cent.

Figures 2.2–2.5 demonstrate the linkages between CO_2 emissions and the relative factors: population, GDP, primary energy consumption and fossil energy consumption. CO_2 emissions fell precipitously in 1989 with the opening of the Hoa-Binh hydropower plant. Vietnam's energy consumption per capita is considered low compared with that for other countries in Asia and the world (100 koe of primary energy in the late 1980s and about 150 koe in recent years). Population and economic growth are not responsible for this noticeable shift, which is due to the increase of hydropower generation, in Vietnam's energy sector. However, apart from this singular event related to the Hoa-Binh plant, population and economic growth are closely correlated with CO_2 emissions. As illustrated in Figure 2.2, CO_2 emissions and population have both been increasing over the decades. Figure 2.3 depicts

Table 2.2 *Factors relative to CO_2 emissions in Vietnam and changes in the factors*

	Factors relative to CO_2 emissions					Changes in the factors over years				
	P	*G* (GDP)	E_0	*E*	*C* (CO_2)	*P*	*G* (GDP)	E_0	*E*	*C* (CO_2)
Unit	million	1000 mill US$(1989)	million toe	million toe	million tCO_2	%	%	%	%	%
	Population	Gross Domestic Product*	Primary Energy Consum.	Fossil Energy Consum.	CO_2 Emissions	Population	Gross Domestic Product	Primary Energy Consum.	Fossil Energy Consum.	CO_2 Emissions
1980	53.7	4.59	5.07	4.44	16.33					
1981	54.9	4.85	4.95	4.38	16.11	2.24	5.56	−2.38	−1.43	−1.39
1982	56.2	5.27	4.95	4.39	16.21	2.26	8.80	0.10	0.32	0.63
1983	57.4	5.65	5.30	4.79	17.73	2.14	7.20	6.95	9.17	9.37
1984	58.7	6.12	5.48	4.92	18.22	2.23	8.30	3.46	2.61	2.81
1985	59.9	6.34	5.45	4.87	18.02	2.08	3.56	−0.65	−1.00	−1.14
1986	61.1	6.49	5.78	5.26	19.46	2.07	2.41	6.15	8.08	8.00
1987	62.5	6.73	6.45	6.01	22.15	2.20	3.64	11.63	14.20	13.83
1988	63.7	7.13	6.79	6.30	23.03	2.04	5.98	5.26	4.73	3.98
1989	64.8	7.47	5.93	4.82	16.81	1.64	4.69	−12.75	−23.39	−27.01
1990	66.2	7.85	6.90	5.46	18.61	2.25	5.10	16.39	13.17	10.68
1991	67.8	8.31	6.63	5.28	17.47	2.33	5.96	−3.84	−3.29	−6.09
1992	69.4	9.03	7.53	6.05	19.96	2.41	8.65	13.56	14.62	14.23
1993	71.0	9.76	8.80	7.24	23.38	2.33	8.07	16.85	19.65	17.10
1994	72.5	10.63	10.10	8.02	25.78	2.09	8.84	14.76	10.82	10.30
1995	74.0	11.64	11.34	7.75	26.41	2.00	9.54	12.30	−3.42	2.44
1996	75.4	12.73	13.54	9.40	32.12	1.88	9.34	19.37	21.34	21.61
1997	76.7	13.85	12.19	9.23	29.50	1.80	8.80	−10.00	−1.85	−8.15

Note: * Converted from GDP in Vietnamese Dong at 1989 prices. Exchange Rate in 1989: 1 US$ = 3763 VND.
Source: General Statistics Office and the author's calculations.

the linkage between GDP and CO_2 emissions. The emissions have been going up with GDP as energy consumption has increased. The movement in CO_2 emissions has followed the movements in primary energy consumption quite closely apart from the increase in hydropower in 1989 (Figures 2.3 and 2.4).

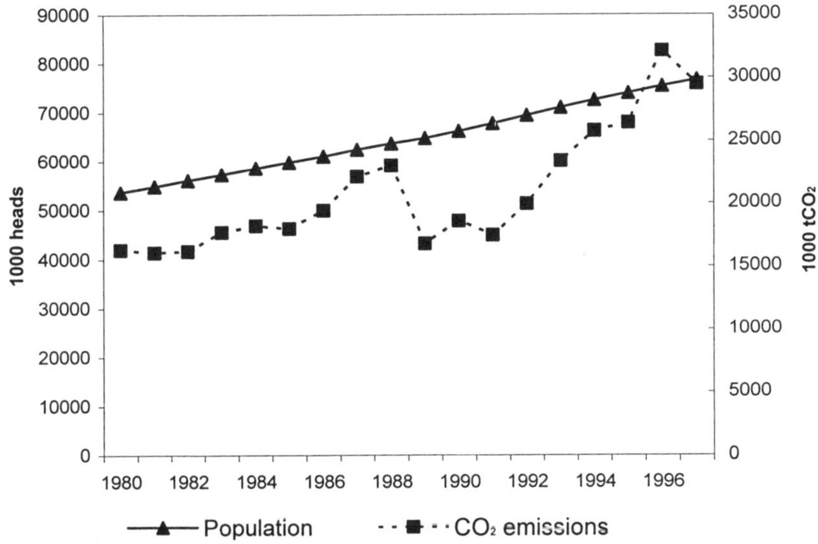

Figure 2.2 Movements of population and CO_2 emissions

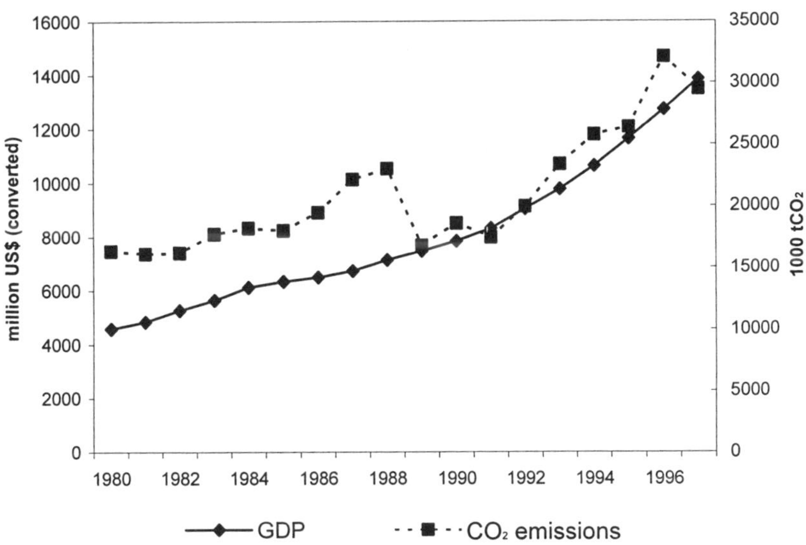

Figure 2.3 Movements of GDP and CO_2 emissions

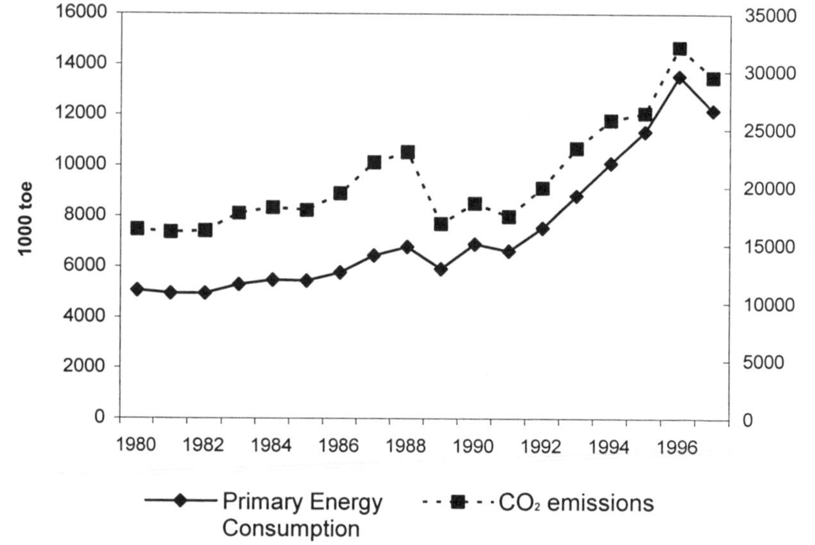

Figure 2.4 Movements of primary energy consumption and CO_2 emissions

The changes in CO_2 emissions are very closely correlated with the changes in fossil energy consumption as shown in Figure 2.5. There is a slight change because of the increased share of oil consumption and reduced share of coal, a higher carbon fuel, in the 1990s.

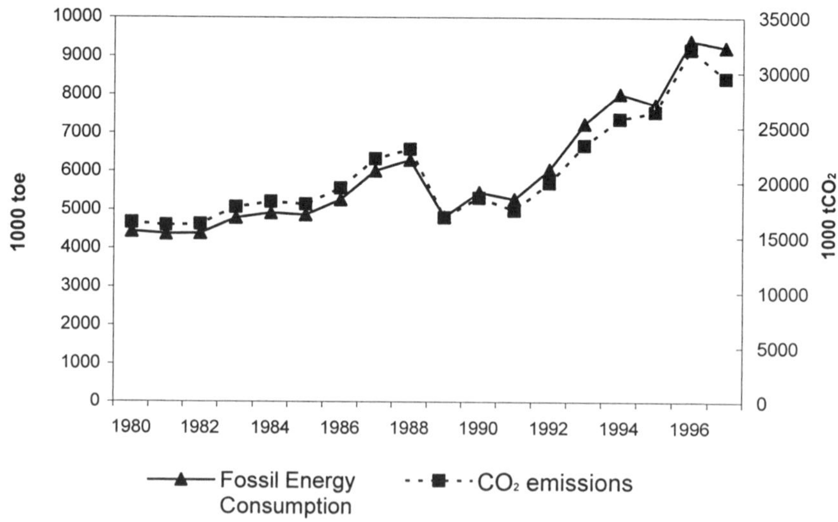

Figure 2.5 Movements of fossil energy consumption and CO_2 emissions

3.2 Movements in the Affecting Factors

Factors affecting the CO_2 emissions from energy consumption considered in the PGESC model include population (*p*), economic intensity (*g*), energy intensity (*e*), fossil intensity (*s*), and carbon intensity (*c*). Table 2.3 shows the values of the factors and how they have changed over time in Vietnam.

The PGESC model estimates the change in economic performance, GDP per capita, which is the leading factor causing change in CO_2 emissions. Factor *g* concerning economic growth has been rapidly increasing over the years. The next most important factor is population, *p*, which has been growing at a rather stable rate of around 2 per cent annually.

Table 2.3 and Figure 2.6 also indicate an irregular change in energy intensity (factor *e*) in Vietnam. As the economic reforms and economic restructuring took place in Vietnam, the energy intensity was sharply curtailed in the early 1980s. It later maintained a stable value of around 0.90 koe/US$ (1989 converted) in the late 1980s and early 1990s. However, the energy intensity has started to increase in recent years.

The increase in hydro-energy as part of primary energy has delayed the increase in CO_2 emissions since 1990. As indicated in the graph, the fossil ratio (factor *s*) has declined significantly over the last decade. Carbon intensity (factor *c*) remained stable in the 1980s, but then fell in the 1990s because of the increase in oil consumption which slightly changed the structure of fossil energy consumption. Carbon intensity in Vietnam has gone up in recent years due to the increase in coal consumption (although it was still low as compared with the corresponding levels in the 1980s).

Figures 2.7–2.11 further illustrate the changes in the affecting factors in Vietnam over the period under consideration. While the columns represent the values of the factors, the lines depict the changes in these factors compared to their values in 1980.

To illustrate the impact of each factor on the change in CO_2 emissions, Figure 2.12 displays the changes comparatively. Factor *g* (economic intensity) has obviously played a significant role in the increasing changes in CO_2 emissions in Vietnam. In addition, factor *p* (population) has also had a part to play in the steady increase in CO_2 emissions. However, other factors (energy intensity, fossil intensity and carbon intensity), particularly fossil intensity, have been decreasing over the period and have continued to be largely responsible for a reduction in CO_2 emissions in the country.

Table 2.3 Factors affecting CO_2 emissions in Vietnam and changes in the factors

	Factors affecting CO_2 emissions					Changes in the affecting factors				
	p	*g*	*E*	*s*	*c*	*p*	*g*	*e*	*s*	*c*
Unit	million	US$(1989) per head*	koe/US$	toe/toe	tCO_2/toe	%	%	%	%	%
	Population	Economic Intensity	Energy Intensity	Fossil Intensity	Carbon Intensity	Change in Population	Change in Economic Intensity	Change in Energy Intensity	Change in Fossil Intensity	Change in Carbon Intensity
1980	53.7	85.5	1.10	0.88	3.68	0.00	0.00	0.00	0.00	0.00
1981	54.9	88.2	1.02	0.88	3.68	2.24	3.24	−7.51	0.97	0.03
1982	56.2	93.9	0.94	0.89	3.69	4.56	9.84	−14.91	1.19	0.35
1983	57.4	98.5	0.94	0.90	3.70	6.80	15.28	−15.11	3.30	0.53
1984	58.7	104.4	0.90	0.90	3.70	9.18	22.12	−18.91	2.46	0.72
1985	59.9	105.9	0.86	0.89	3.70	11.45	23.89	−22.21	2.10	0.58
1986	61.1	106.2	0.89	0.91	3.70	13.75	24.31	−19.36	3.96	0.50
1987	62.5	107.7	0.96	0.93	3.68	16.25	26.06	−13.15	6.36	0.18
1988	63.7	111.9	0.95	0.93	3.66	18.62	30.93	−13.74	5.82	−0.54
1989	64.8	115.3	0.79	0.81	3.49	20.57	34.85	−28.11	−7.09	−5.24
1990	66.2	118.5	0.88	0.79	3.41	23.29	38.61	−20.39	−9.65	−7.32
1991	67.8	122.7	0.80	0.80	3.31	26.16	43.53	−27.75	−9.14	−9.99
1992	69.4	130.1	0.83	0.80	3.30	29.19	52.28	−24.48	−8.29	−10.30
1993	71.0	137.4	0.90	0.82	3.23	32.21	60.81	−18.35	−6.10	−12.21
1994	72.5	146.5	0.95	0.79	3.21	34.97	71.45	−13.90	−9.33	−12.62
1995	74.0	157.4	0.97	0.68	3.41	37.68	84.11	−11.74	−22.02	−7.32
1996	75.4	168.9	1.06	0.69	3.42	40.27	97.59	−3.64	−20.73	−7.11
1997	76.7	180.5	0.88	0.76	3.20	42.79	111.18	−20.29	−13.55	−13.08

Note: * Converted from GDP in Vietnamese Dong at 1989 prices. Exchange Rate in 1989: 1 US$ = 3763 VND. Changes compared with 1980.

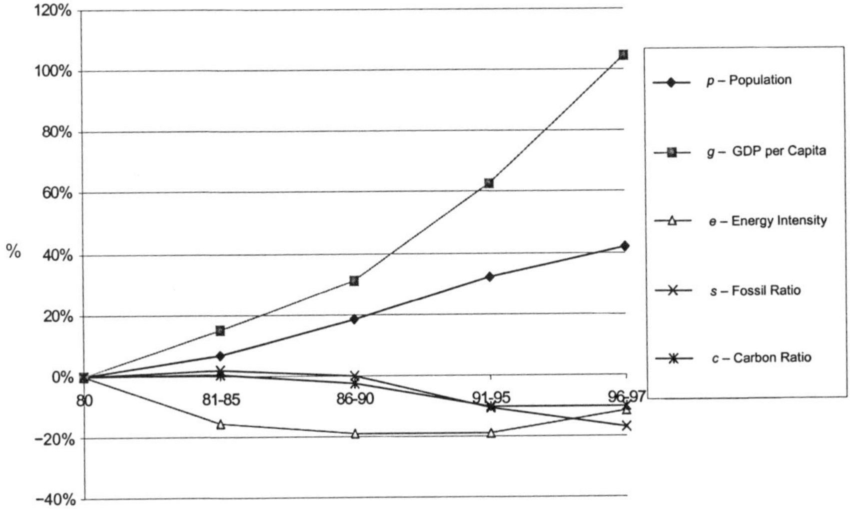

Figure 2.6 Movements in the affecting factors over periods

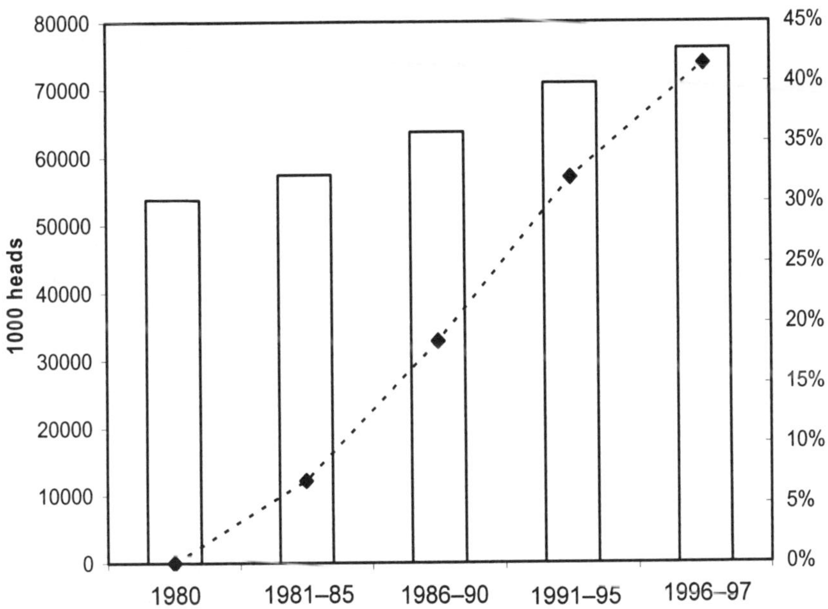

Figure 2.7 Changes in factor p *– population*

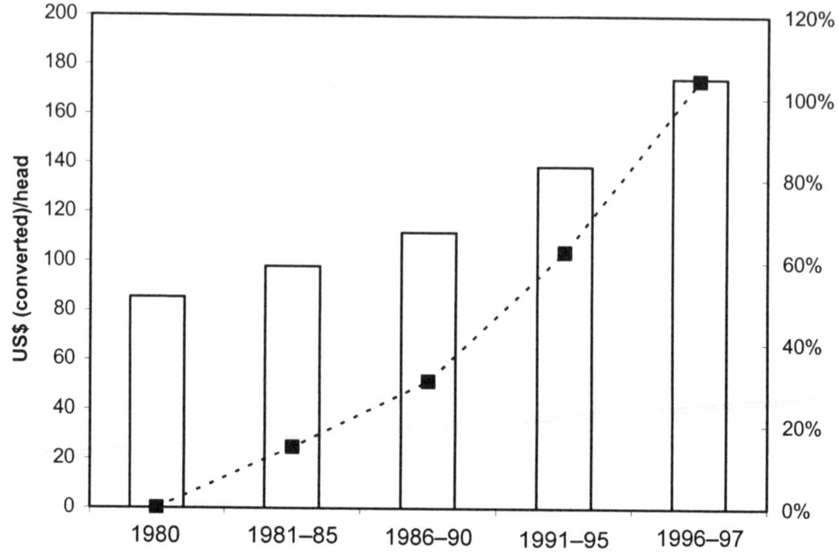

Figure 2.8 Changes in factor g *– economic intensity*

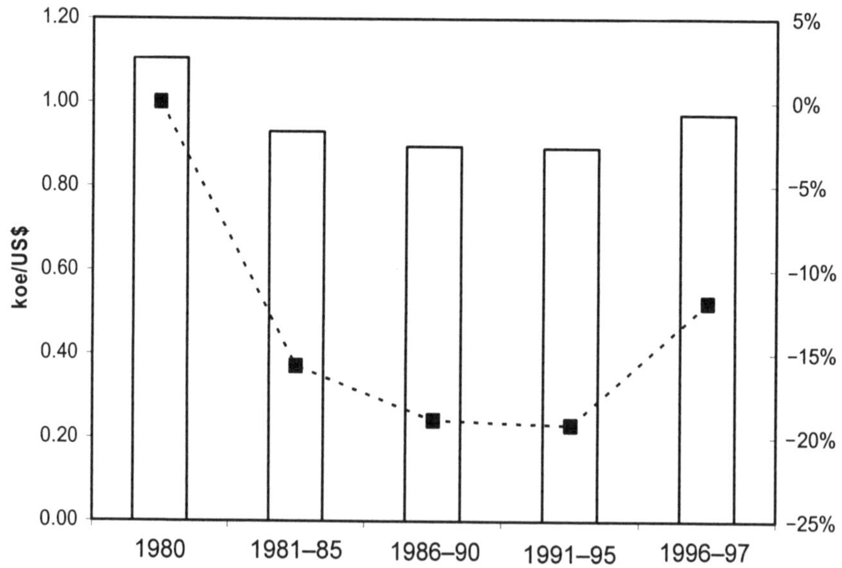

Figure 2.9 Changes in factor e *– energy intensity*

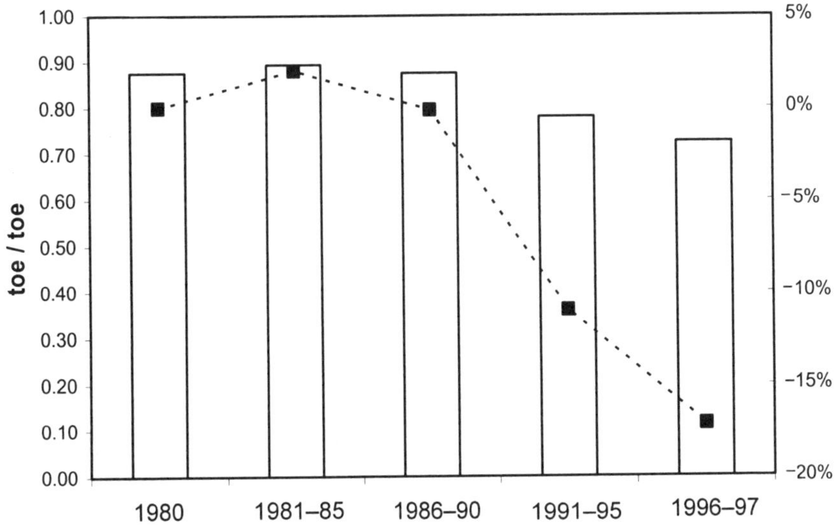

Figure 2.10 Changes in factor s *– fossil intensity*

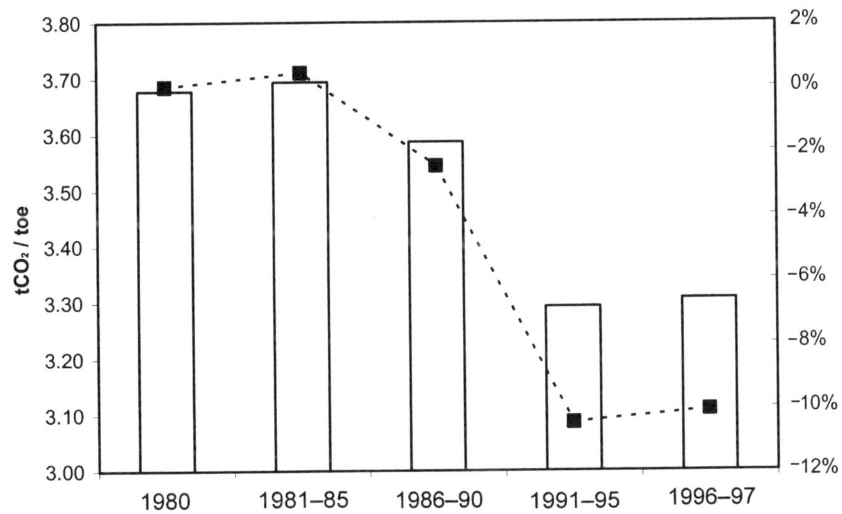

Figure 2.11 Changes in factor c *– carbon intensity*

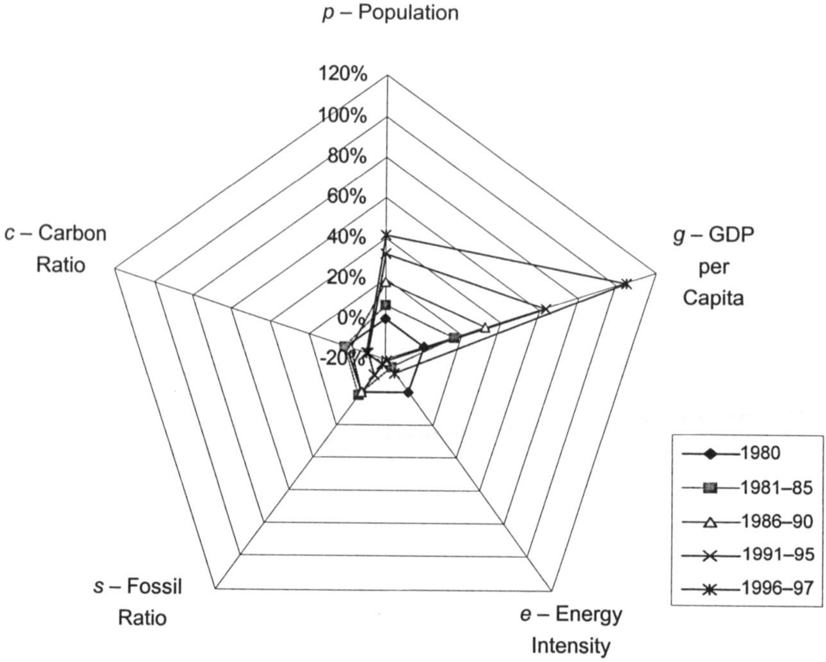

Figure 2.12 Movements of the affecting factors over periods

4 TRENDS IN CO_2 EMISSIONS IN VIETNAM

This section considers the trends in CO_2 emissions due to energy consumption in Vietnam. Based on the three growth scenarios for the Vietnamese economy, three scenarios for CO_2 emissions in this country have been projected.

4.1 Assumptions for Economic Growth, Population and Energy Consumption

This study assumes that the Vietnam economy will perform in accordance with the three scenarios for economic development to the year 2010 that have been drawn up by the Vietnamese Government. As shown in Table 2.4, the economy is predicted to grow at annual rates of 8.2 per cent, 7.2 per cent, and 6.5 per cent during 1997–2005, and at annual rates of 7.8 per cent, 7.2 per cent, and 6.5 per cent during 2006–2010 based on high, medium, and low economic development scenarios, respectively.

The PGESC model predicts that economic growth of about 7.2 per cent annually (in the medium scenario) will result in an annual growth rate of primary energy consumption of about 10 per cent (Table 2.6). Rising population levels (Table 2.5) and continued economic growth are the major factors giving rise to the increase in CO_2 emissions in Vietnam.

Table 2.4 Forecasts for economic growth to 2010

	% per year	1997–2005	2006–2010
Low Scenario	Average Growth	6.5	6.5
Base Scenario	Average Growth	7.2	7.2
High Scenario	Average Growth	8.2	7.8

Source: Based on Institute for Development Strategy (Vietnam, 1999), National Prog. on Energy KHCN (09.02.1999) and the author's assumption.

Table 2.5 Forecasts for population to 2010

		2000	2005	2010
Population	million heads	81	87–88	94
Annual Growth	1997–2005, %		1.80	
Annual Growth	2006–2010, %		1.50	

Source: Based on Institute for Development Strategy (Vietnam, 1999) and the author's assumption.

4.2 Trends in CO_2 Emissions

Owing to the rapid increase in energy consumption, CO_2 emissions will increase significantly over the coming decade. Table 2.7 indicates energy consumption and CO_2 emissions in Vietnam for each growth scenario (which are illustrated in Table 2.6). Primary energy consumption in Vietnam based on the medium scenario will be 24.3 million toe in 2005, and will reach 37 million toe in 2010. This is 2.7 times the consumption in 1997. Fossil energy consumption will be 16 million toe in 2005 and nearly 25 million toe in 2010. CO_2 emissions will be 51.5 million tons of CO_2 in 2005 and nearly 75 million tons of CO_2 in 2010. Thus, by 2010, Vietnam's CO_2 emissions are predicted to increase to more than 3.5 times their 1990 levels.

Table 2.6 Assumptions for energy intensity based on economic growth scenarios

Medium Economic Growth Scenario				
	1986–1990	1991–1997	1997–2005	2005–2010
Average Growth of GDP, %	3.90	8.40	7.20	7.20
Average Growth of Primary Energy Consumption, %	5.00	11.20	10.08	10.44
Energy Elasticity Coefficient	1.28	1.33	1.40	1.45
Average Growth of Energy Intensity, %	0.92	1.76	1.83	2.38

High Economic Growth Scenario				
	1986–1990	1991–1997	1997–2005	2005–2010
Average Growth of GDP, %	3.90	8.40	8.20	7.80
Average Growth of Primary Energy Consumption, %	5.00	11.20	11.48	11.31
Energy Elasticity Coefficient	1.28	1.33	1.40	1.45
Average Growth of Energy Intensity, %	0.92	1.76	1.98	2.53

Low Economic Growth Scenario				
	1986–1990	1991–1997	1997–2005	2005–2010
Average Growth of GDP, %	3.90	8.40	6.50	6.50
Average Growth of Primary Energy Consumption, %	5.00	11.20	9.10	9.43
Energy Elasticity Coefficient	1.28	1.33	1.40	1.45
Average Growth of Energy Intensity, %	0.92	1.76	1.71	2.21

Figure 2.13 and Figure 2.14 illustrate CO_2 emissions from energy consumption in Vietnam up to 2010 for each scenario. For the low growth scenario, CO_2 emissions are expected to reach 49 million tons by 2005 and 68.9 million tons by 2010. For the high growth scenario, CO_2 emissions are expected to reach 54.7 million tons by 2005 and 81.4 million tons by 2010. In all the scenarios, Vietnam will produce over three times the CO_2 emissions it produced in 1990. Although CO_2 emissions are predicted to rise rapidly in Vietnam, these values are still low in comparison with other countries.

Table 2.7 Energy consumption and CO_2 emissions based on growth scenarios

compared with 1997

		In value			%		
		1997	2005	2010	1997	2005	2010
Primary Energy Consumption, mill toe	Medium Growth	13.46	24.31	37.00	100.0	180.6	274.9
	High Growth	13.46	25.82	40.42	100.0	191.8	300.3
	Low Growth	13.46	23.25	34.22	100.0	172.8	254.2
Fossil Energy Consumption, mill toe	Medium Growth	9.77	16.24	24.72	100.0	166.2	253.0
	High Growth	9.77	17.24	27.00	100.0	176.5	276.4
	Low Growth	9.77	15.53	22.85	100.0	159.0	233.9
CO_2 Emissions mill tCO_2	Medium Growth	32.31	51.56	74.54	100.0	159.6	230.7
	High Growth	32.31	54.74	81.43	100.0	169.4	252.0
	Low Growth	32.31	49.31	68.93	100.0	152.6	213.3

Source: The author's assumption and calculations.

Together with the changes in CO_2 emissions, Figure 2.14 also presents the changes in primary energy consumption and the changes in fossil energy consumption in Vietnam. Compared with the emissions in 1997, CO_2 emissions will be more than one and a half times that level by 2005 and will more than double by 2010.

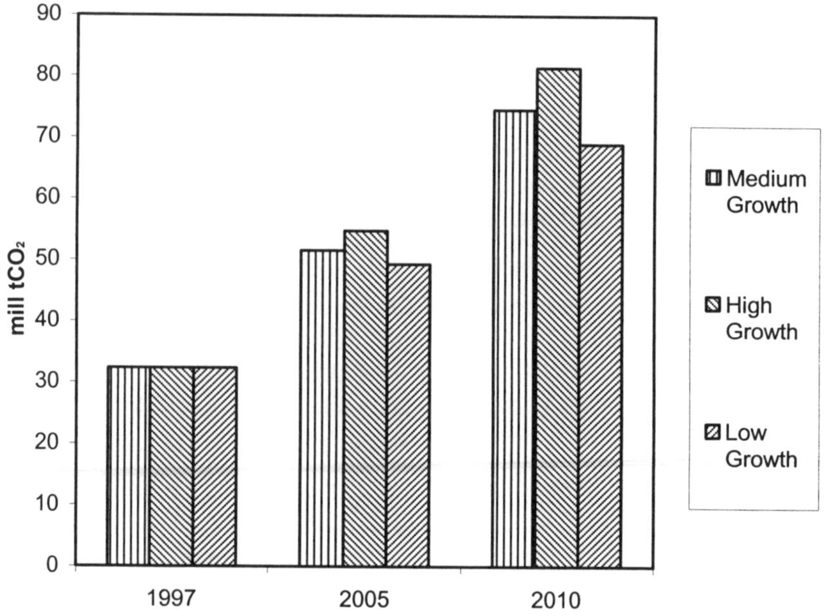

Figure 2.13 CO_2 *emissions in growth scenarios*

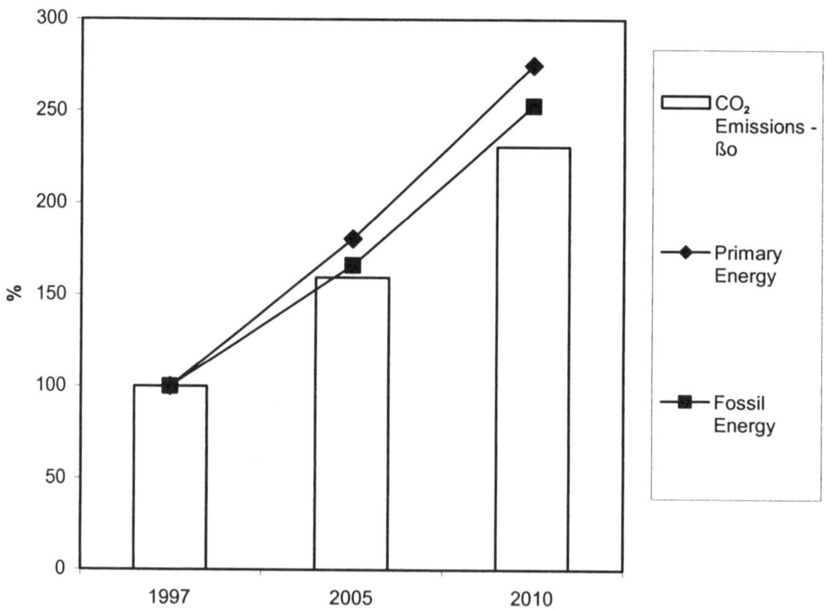

Figure 2.14 *Changes of energy consumption and* CO_2 *emissions in baseline scenario (%)*

5 CONCLUSION

The PGESC model examines five critical factors that potentially affect CO_2 emissions: population growth, GDP per capita, energy intensity, fossil fuel intensity and carbon intensity. Each of these factors is examined for Vietnam. The model predicts that the most important factor is the growth of GDP per capita. Population is the next most important factor. Historically, these two factors explain a great deal of the increase in carbon emissions over the last 20 years. The only major exception to this pattern is a sharp decline in emissions associated with the 1989 opening of the Hoa-Binh hydroelectric facility. With the future population and economic growth predicted for Vietnam, CO_2 emissions are expected to increase sharply over the next decade. CO_2 emissions will reach 69–81 million tons by 2010, which is almost a fourfold increase over 1990 levels.

CO_2 abatement is not easy to implement in developing countries. The countries often face a shortage of capital investment to carry out their development plans and strategies. Vietnam could reduce CO_2 emissions by encouraging the implementation of flexible mechanisms as described in the Kyoto Protocol, but it will have to weigh the cost of CO_2 abatement against other pressing social and economic concerns.

ACKNOWLEDGEMENTS

I am grateful to the International Development Research Center of Canada (IDRC) and the Economy and Environment Program for Southeast Asia (EEPSEA) for funding this research. To complete this chapter, I would like to express my sincerest thanks to Dr David James, Dr Mohan Munasinghe, and Dr Nguyen Minh Due for very useful comments. I would especially like to convey my wholehearted thanks to Dr Ching-Cheng Chang, Dr Daigee Shaw, Dr Robert Mendelsohn and the reviewers of earlier versions of this chapter for their very valuable comments and significant encouragement. My thanks are further expressed to the Institute of Economics (IEAS, Academia Sinica of Taiwan) for meaningful and substantial support.

NOTES

1. This chapter is based on the research project 'Assessment and Forecast of CO_2 Emissions in Vietnam', funded by the International Development Research Center (IDRC, Canada).
2. It is noted that this research began in late 2000, before the United States under President Bush stated the US's position in the Kyoto Protocol.
3. Primarily for two reasons: first, their priorities are economic growth and poverty reduction; and secondly, industrialized countries consume far more energy and thus produce

more greenhouse gases.

4. Due to the lack of sufficient and reliable data on the consumption of non-commercial energy in Vietnam such as fuel wood, this research work focuses only on the consumption of commercial energy in the country.
5. Ehrlich and Ehrlich (1990).
6. Energy consumption is generally measured by tons of oil equivalent (toe). It can be also measured by barrels of oil equivalent (boe) or British thermal unit (Btu). In this study, the energy unit used is toe.
7. Source: Manuals on energy units.
8. Here tC is tons of carbon content; 44 and 12 are the molecular weights of carbon dioxide (CO_2) and carbon (C), respectively ($CO_2 = 12 + 2 \times 16 = 44$ and $C = 12$).
9. 'Kaya-Identity' on carbon dioxide emissions: $CO_2 = CO_2/E \times E/\text{GDP} \times \text{GDP}/P \times P$
CO_2 Emissions = Carbon Intensity $\times$ Energy Intensity $\times$ GDP per Capita $\times$ Population
10. c^0 is not mentioned here because it is constant, being 12/44.
11. For example, The Intergovernmental Panel for Climate Change (IPCC), The Center on Energy and Environment of the UN Environmental Program, the US Department of Energy, Berkeley Energy Lab (US), the Stockholm Environment Institute in Boston, the CO_2 Analysis Center (US), the International Energy Agency (IEA), and the Environmental Research Institute (Japan).
12. The emissions coefficients used are averaged from the values provided by the Intergovernmental Panel for Climate Change and the Stockholm Environment Institute in Boston.
13. Source: World Bank (2000), *The Little Data Book 2000*, from the World Development Indicators.

REFERENCES

Ehrlich, P.R. and A.H. Ehrlich (1990), *The Population Explosion*, New York: Simon and Schuster.

IISD (International Institute for Sustainable Development) (1998), *A Guide to Kyoto: Climate Change and What it Means to Canadians*.

UNDP and WRI (World Resources Institute) (1999), *Promoting Development while Limiting GHG Emissions – Trends and Baselines*, edited by Jose Goldemberg and Walter Reid.

World Bank (1998), *Issues and Options in Energy and Environment*.

World Bank (2000), *The Little Data Book 2000*, from the World Development Indicators.

Zhang, ZhongXiang (1999), 'Is China taking Actions to Limit its Greenhouse Gas Emissions? Past Evidence and Future Prospects', in J. Goldemberg and W. Reid (eds), *Promoting Development while Slowing Greenhouse Gas Emissions Growth*, United Nations Development Programme, New York, pp. 45–57.

3. Baseline forecasting for greenhouse gas reductions in Taiwan: a dynamic CGE analysis

Ping-Cheng Li, Shih-Hsun Hsu, Chung-Huang Huang and Hsing-Hua Lin

1 INTRODUCTION

The adoption of the Kyoto Protocol in December 1997 by the COP3 (Third Conference of the Parties) to the UNFCCC (United Nations Framework Convention on Climate Change) sets the timetable and reduction targets of greenhouse gas emissions for the developed countries. To avoid the possibility of international trade sanctions in the future, it is important for Taiwan to take on some of the obligations laid down in the Kyoto Protocol, even though there is no abatement responsibility specified by the Kyoto Protocol for the developing countries. In order to fulfill her abatement responsibility, it is essential that Taiwan accurately forecast her carbon dioxide (CO_2) baseline. Without a reasonable baseline projection, it is not possible to perform any economic evaluation of emissions reduction strategies.

According to the conclusions reached at the 1998 National Energy Conference, Taiwan proposes to reduce her greenhouse gas (GHG) emissions to about 10 per cent below or above her 2000 level for the commitment year of 2020. However, there has been much debate in relation to the baseline projection generated by the Taiwan-MARKAL model.[1] The distinction between a 'top-down' and 'bottom-up' approach to modeling the energy sector has received a great deal of attention in discussions on projected emissions growth in Taiwan. One of the major conclusions of the Conference has been to improve Taiwan's CO_2 baseline forecasting by making use of both the top-down and bottom-up approaches. The model adopted in this chapter is the TAIGEM©-D (TAIwan General Equilibrium Model-Dynamic), a dynamic top-down CGE model.

The baseline (or reference case) provides growth projections in relation to carbon dioxide emissions, labor and capital in each country or region, and the associated changes throughout the rest of the economy in the absence of any policy measures. The results of policy simulations are then interpreted as deviations from the baseline. For example, the influence of an emissions reduction policy can be isolated from other economic trends by comparing emissions growth in the simulation against emissions in the baseline. The effect on emissions may also be reported as a percentage reduction from the baseline emissions projection.

There are a number of factors which might affect baseline projections, for example, the differences in the models used, the degree of substitution between factors of production, the adoption of the new energy-saving technologies and differences among market mechanisms. Moreover, the baseline projections are also affected by different assumptions with regard to the future values of the exogenous variables, such as the GDP growth rate, population growth rate, rate of technical progress, and so on.

The analysis of the GHG emissions baseline in this chapter is based on the forecasting results from TAIGEM©-D. TAIGEM©-D is a dynamic, multi-sectoral, computable general equilibrium (CGE) model of the Taiwan economy, developed specifically to analyse climate change response issues. TAIGEM©-D is derived from the Australian ORANI and MONASH models (Dixon et al., 1997; 1998; Dixon and Rimmer, 2000). Those most significant features that distinguish TAIGEM©-D and MONASH are the coverage of GHG emissions and the inclusion of inter-fuel substitution and technology bundles. We use historical simulations to generate up-to-date data for our baseline forecasting. The results from the TAIGEM©-D model and other models (for example, the Taiwan-MARKAL energy model) are compared and analysed.

The next section provides an overview of the TAIGEM©-D model. The main features such as the dynamic mechanism, technology bundles, and so on, are given special emphasis. Section 3 focuses on the major assumptions and baseline forecasting results of the TAIGEM©-D model. Section 4 compares the results of TAIGEM©-D with the Taiwan-MARKAL energy model. The final section provides concluding comments.

2 AN OVERVIEW OF THE TAIGEM DYNAMIC GENERAL EQUILIBRIUM MODEL

TAIGEM©-D is descended from the TAIGEM© model, developed specifically to analyse climate change issues, such as baseline forecasting and climate change response policies. As mentioned above, TAIGEM© is a multi-sectoral, computable general equilibrium (CGE) model of the Taiwan economy derived from the ORANI model (Dixon et al., 1982). The input–output database was compiled from the 150-sector Use Table of the

1994 input–output tables for Taiwan. TAIGEM© distinguishes 160 sectors, 6 types of labor, 8 types of margins and 170 commodities. Like ORANI, TAIGEM© was designed for comparative-statics, that is for projecting what difference a shock would make to the economy at a point in time. The most significant features that distinguish TAIGEM©-D from TAIGEM© are the inclusion of inter-fuel substitution, technology bundles and a dynamic mechanism capable of projecting the development of the economy through time. With TAIGEM©-D, we have derived annual projections of CO_2 emissions, the GDP growth rate, and other economic variables.

2.1 Dynamic Mechanism

According to Dixon and Parmenter (1996), dynamic mechanisms of the CGE model may be categorized into four broad cases, namely, (1) exogenous investment, a recursive model, (2) endogenous investment, but still a recursive model, (3) a non-recursive multi-period model, and (4) a non-recursive multi-period model with optimizing investment behavior.

In Case 1, investment is exogenous. In Case 2, investment and capital accumulation in year $t+1$ depend on expected rates of return for year $t+2$, which we assume are determined by actual returns to and costs of capital in year $t+1$. In both Cases 1 and 2, the models are recursive, that is they can be solved for year 1 and then for year 2 and so on. In Case 3, expected rates of return for year $t+2$ are assumed equal to the actual rates of return for year $t+2$, namely, expectations are rational or consistent. In Case 4, the behavior of investors is explicitly optimizing. Relative to the recursive models in Cases 1 and 2, the respective solutions for the Case 3 and Case 4 models require a more sophisticated computational approach for handling the computations for all of the years simultaneously.

In our TAIGEM©-D forecasting and policy simulations, we solve a large (160 industries) recursive model incorporating externally supplied, realistic macro-forecasts. That is, our approach is an application of Case 2. A dynamic model such as TAIGEM©-D is beneficial when analysing climate change policies since the timing of policy implementation and the adjustment path an economy follows are highly relevant in the climate change policy debate.

2.2 Structure of Production: Non-Electricity Sectors

TAIGEM©-D allows each industry to produce several commodities, using as inputs domestic and imported commodities, several types of labor, land, capital, several types of energy and 'other costs'. In addition, commodities destined for export are distinguished from those for local use. The multi-input, multi-output production specification is kept manageable by a

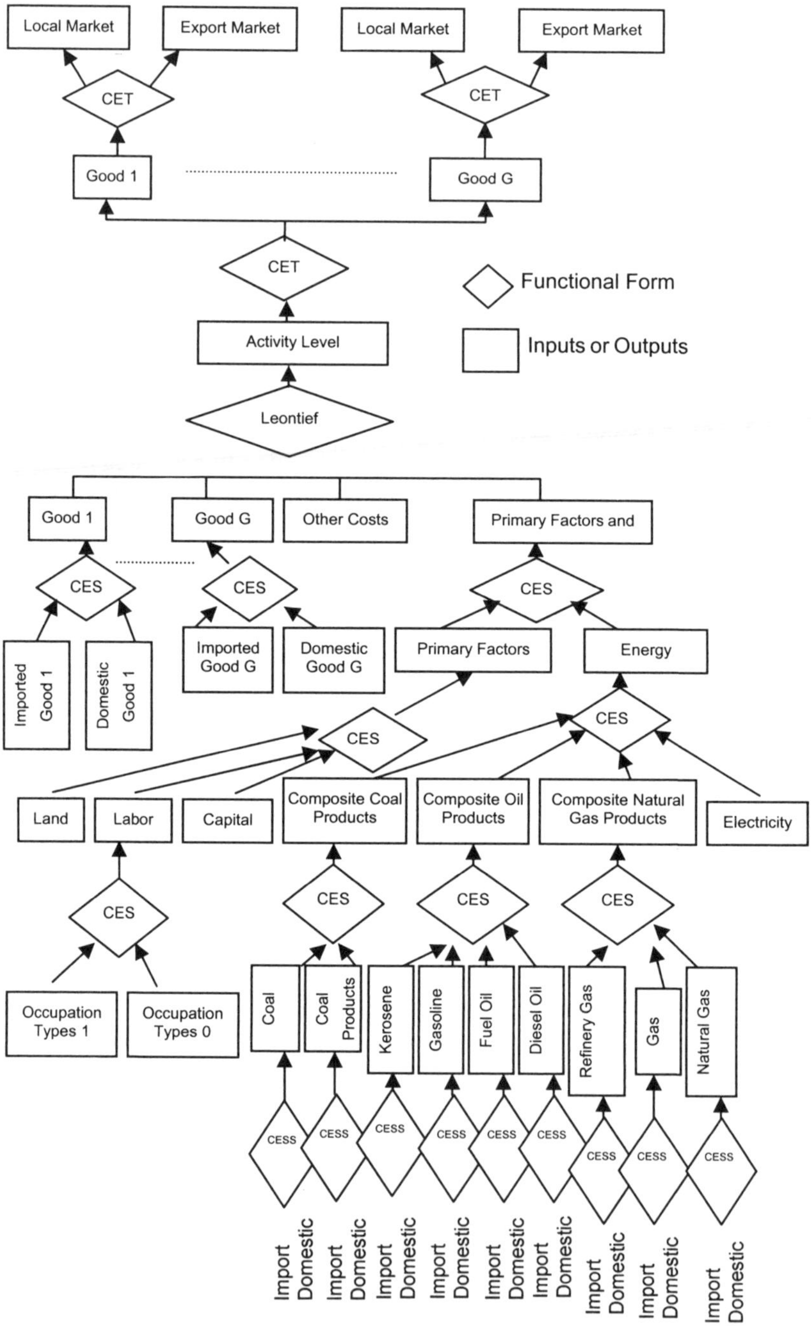

Figure 3.1 Structure of production: non-electricity sectors

series of separability assumptions, illustrated by the nesting shown in Figure 3.1, which depicts the production structure of the non-electricity sectors of the TAIGEM©-D model.

The input demand for industrial production is formulated by a five-level nested structure, with the production decision-making process being independent at each level. Assuming cost minimization and a technology constraint at each level of production, producers will make optimal input-demand decisions. At the top level, commodity composites and a primary-factor composite are combined using a Leontief production function. Consequently, they are all demanded in direct proportion to the industrial activity. At the second level, each commodity composite is a CES (constant elasticity of substitution) production function of domestic goods and the imported equivalent (the Armington assumption). Energy and primary-factor composites are a CES aggregation of energy composites and primary-factor composites.

At the third level, the primary-factor composite is a CES aggregation of labor, land and capital and the energy composite is a CES aggregation of coal products composites, oil products composites, natural gas products composites and electricity. At the fourth level, the labor composite is a CES aggregation of managers, professional specialists, white collar workers, technical workers and unskilled workers; the coal products composite is a CES aggregation of coal and coal products; the oil products composite is a CES aggregation of gasoline, diesel oil, fuel oil and kerosene; the natural gas products composite is a CES aggregation of refinery gas, gas and natural gas. At the bottom level, the energy composite is a CES aggregation of domestic goods and imported goods.

Like the ORANI model, the output structure of TAIGEM©-D allows each industry to produce a mixture of all the commodities. Moreover, conversion of an undifferentiated commodity into goods destined for export and local use is governed by a CET (constant elasticity of transformation) transformation frontier.

2.3 Technology Bundles in the Electricity Sector

In TAIGEM©-D, production in the electricity sector is modeled using the 'technology bundle' approach derived from the Australian ORANI-E model (McDougall, 1993a, 1993b) and the MEGABARE (ABARE, 1996) model. With this approach, electricity can be generated from coal, petroleum, gas, nuclear power, hydro-based or renewable technologies. The electricity industry is able to substitute between technologies in response to changes in their relative costs. By modeling energy-intensive industries in this way, TAIGEM©-D restricts substitution to known technologies, thereby preventing technically infeasible combinations of inputs being chosen as model solutions. While retaining the extensive interaction with other sectors of the economy obtained in 'top down' models, TAIGEM©-D moves further toward the realism of the 'bottom up' approach.

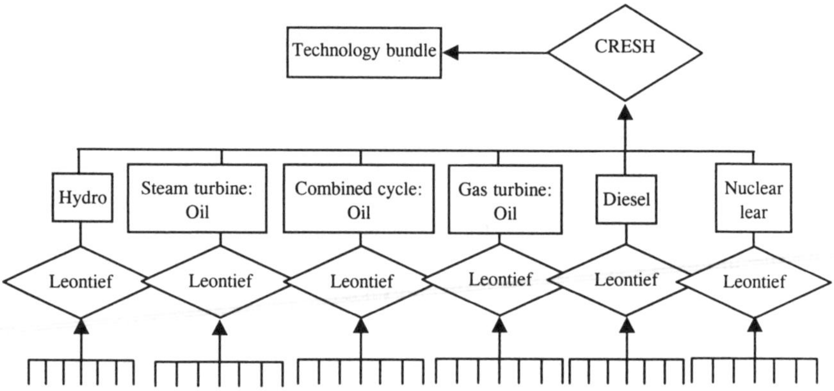

Figure 3.2 Technology bundle of the TAIGEM©-D model: electricity sector

The way in which the technology bundle approach ensures that the pattern of input use is consistent with known technologies is illustrated in Figure 3.2. In the TAIGEM©-D model, 10 known technologies are used to generate electricity, namely, the hydro, steam turbine-oil, steam turbine-coal, steam turbine-gas, combined cycle-oil, combined cycle-gas, gas turbine-oil, gas turbine-gas, diesel and nuclear technologies. All electricity generated by these technologies is transferred to the end-use electricity sector.

The output of the electricity sector consists of a CRESH[2] aggregate for each electricity technology and this technology requires fixed proportions of intermediate inputs, with the exception of energy inputs and primary factors.

3 BASELINE PROJECTION RESULTS FROM TAIGEM©-D

Like MONASH, three types of simulations are made routinely with the TAIGEM©-D model. The first is the historical simulation that we use to generate up-to-date data for our forecasting simulation. Since models designed for forecasting contain dynamic equations that require initial conditions from the base year, forecasts can be rather sensitive to these initial conditions. Moreover, through the historical simulations, we can calibrate detailed patterns of changes in technology and household tastes over the historical period.

The second is the forecasting simulation that is designed so that we might incorporate into the forecasts as much specialist information as is available, allowing us to project prospects for likely developments in the structure of the economy. The third is the policy simulation that is conducted by projecting deviations from an explicit control path (that is the forecasting simulation), which shows the effects of policy changes or other shocks that are of interest (for example, GHG mitigation policies).

The policy simulations are not conducted in this study since the GHG mitigation strategies are not evaluated. The first two types of simulation that we use are achieved using two different closures of the model, that is through two different selections of exogenous variables. Historical closures include in their exogenous set two types of variables: observables and assignables. Observables are those for which movements can be readily observed from statistical sources. Historical closures vary between applications depending on data availability. When we forecast the baseline, we need the results of technology movements and taste movements that are solved by historical simulation. Thus, in the historical closure, the technology movements and taste movements are endogenous variables.

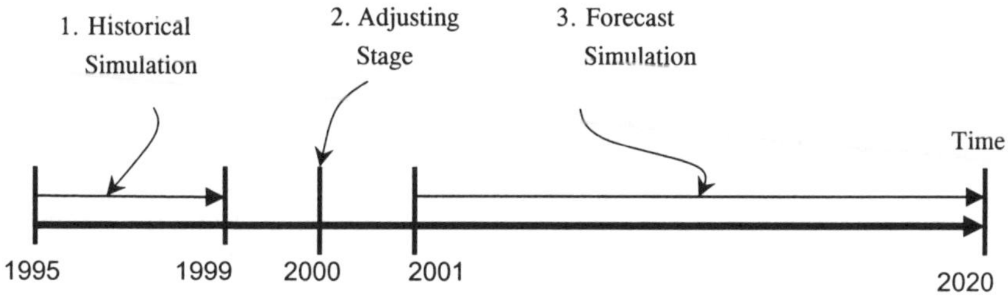

Figure 3.3 Simulation closure of the TAIGEM©-D model

In the forecasting closures, the experts' forecasting values are exogenous variables, such as tradable export prices and quantities, employment and the economic growth rate. These exogenous variables are altered during the forecasting period and the closures must be modified to incorporate the new exogenous variables. If we can obtain the forecasts of more experts, then we can make the results of our forecast more accurate.

For the CO_2 baseline forecasting, we consider the period from 1995 to 2020[3] as shown in Figure 3.3. The initial database of the TAIGEM©-D model consists of the 1994 input–output tables. Three situations are specified as follows:

1. Historical closure during the 1995–99 period: since official data on private consumption, investment, government consumption, exports, the

exchange rate and labor employed are available from the Directorate-General of Budget, Accounting and Statistics (DGBAS), we regard the growth rates of these variables as exogenous.

2. Closure for the year 2000: there are two differences between the closure for the year 2000 and the historical closure for the period 1995–99. First, total labor employed is treated as exogenous in the historical simulation for the 1995–99 period. However, since official data on total labor employed in the year 2000 are not available, we set it as endogenous. Second, the aggregate price index is set as the numeraire in the historical closure for the 1995–99 period, but the consumer price index is set as the numeraire in the closure for the year 2000. Third, the exchange rate is set as exogenous in the historical closure for the 1995–99 periods, but there is a closure swap between the exchange rate and the productivity growth rate of primary factors in the closure for the year 2000.
3. Our aim with TAIGEM©-D was to use all of the information for the final year (that is the year 2000) that was available from the DGBAS, both in published and unpublished form. The results also provide quite detailed estimates of changes in technology over the historical period. We use these as the starting point for devising forecasting simulations in relation to technical change to be incorporated into our baseline forecasts with TAIGEM©-D.
4. Forecast closure in the 2001–20 period: most exogenous variables in the historical closure for the 1995–99 period are set as endogenous in the forecast closure. In the baseline forecast, private consumption, investment, government consumption, exports and imports are determined in the model.

Table 3.1 exhibits the growth rates of major economic variables over the period 1995 to 2020. The historical economic statistics and projected values are taken from DGBAS publications such as 'Quarterly National Economic Trends – Taiwan Area, the Republic of China', 'Commodity-Price Statistics Monthly in the Taiwan Area of the Republic of China', 'Report on the Survey of Family Income and Expenditure in the Taiwan Area of the Republic of China', and 'Monthly Statistics of the Republic of China'. Moreover, data from CEPD (the Council for Economic Planning and Development) are also used, for example, 'Industry of Free China' and 'The Plan for National Development into the Next Century'.

For the historical simulation, for example, the growth rates of exports are set at 12.34, 6.74, 9.08, 2.41, 11.90 and 17.55 per cent for the 1995–2000 periods. The growth rates of investment are given as 7.31, 1.66, 10.65, 8.01, 1.78 and 8.61 per cent. The growth rates of household consumption are set at 5.63, 6.53, 7.26, 6.52, 5.37 and 4.93 per cent. The growth rates of government expenditure are 3.15, 6.66, 5.87, 4.12, −6.49 and 0.55 per cent. The growth rates of the aggregate price index are set at 2.02, 3.11, 1.68, 2.64 and −1.42 per cent for the 1995–99 period. The growth rates of the exchange rate are set at 4.04, 0.73, 18.55, −1.23 and −2.48 per cent. The growth rates of the import price index (c.i.f.) are 9.12, −1.90, 0.11, 2.22 and −2.96 per cent. Finally, the growth rates of the export price index are 6.23, 1.29, 1.24, 3.83 and −6.11 per cent.

As shown in Table 3.1, we assume that the productivity growth rate of primary factors and the CPI growth rate will increase by an annual rate of 2.5 per cent[4] for the years 2000–2020 and by 2 per cent from the year 2000 and beyond, respectively. Table 3.2 shows the baseline scenario assumptions.

Table 3.3 and Figure 3.4 exhibit CO_2 emission results with different assumptions regarding GDP growth rates. In the first case, the GDP growth rates are endogenously determined. The forecasted results show that total CO_2 emissions amount to 386 million tons and fossil fuel-related CO_2 emissions total 361 million tons in the year 2020. Of most interest is the historical simulation from 1995 to 2000 for GDP growth rates solved from TAIGEM©-D. They are very close to the actual values published by the DGBAS. The other case is based on the assumption that GDP growth rates are set exogenously. The growth rates are taken from projections made by Taiwan's Academia Sinica. The results show that in the year 2020 total CO_2 emissions come to 454 million tons and fossil fuel-related CO_2 emissions amount to 426 million tons. Because the exogenous GDP growth rates given by Academia Sinica are higher than the endogenous GDP growth rates generated from TAIGEM©-D and because the total CO_2 emissions are roughly proportional to GDP, the projected CO_2 emissions using the exogenous GDP growth rates exceed those generated by endogenously-determined GDP growth rates.

Table 3.4 illustrates the structure of energy demand and electricity generation based on the TAIGEM©-D baseline projection using endogenously-determined GDP growth rates. As we can see from this table, the percentage share of the demand for coal and natural gas in Taiwan generally increases over time, while the percentage share of the demand for hydro-power and oil decreases. The increase in the percentage share of energy demanded is the largest for coal, followed by natural gas. The demand for nuclear energy changes over time because of the construction of the fourth nuclear power plant and the decommissioning of the first nuclear power plant (see Table 3.2 for the time-frame). Since there will be no further construction projects after the fourth nuclear power plant, the percentage share of nuclear power in Taiwan's energy demand may decease significantly

Table 3.1 Exogenous shocks for forecasting the CO_2 emissions baseline: from 1995 to 2020

Macroeconomic variables growth rate (%)	1995	1996	1997	1998	1999	2000	2001–2020
Energy-saving decline rate	−0.31	−0.16	−0.08	−0.15	−0.6	−0.6	−0.6
Imports	9.78	6.03	13.74	6.31	4.41	14.53	endog*
Household consumption	5.63	6.53	7.26	6.52	5.37	4.93	endog.
Exports	12.34	6.74	9.08	2.41	11.90	17.55	endog.
Investment	7.31	1.66	10.65	8.01	1.78	8.61	endog.
Government expenditure	3.15	6.66	5.87	4.12	−6.49	0.55	endog.
Households**	3.01	3.49	3.02	2.69	2.54	2.30	2
Employment trend	1.19	0.26	1.19	1.24	1.00	endog.	endog.
Aggregate price index	2.02	3.11	1.68	2.64	−1.42	−1.73	endog.
Exchange rate	4.04	0.73	18.55	−1.23	−2.48	−1.34	endog.
Imports price index (c.i.f.)	9.12	−1.90	0.11	2.22	−2.96	endog.	endog.
Exports price index	6.23	1.29	1.24	3.83	−6.11	endog.	endog.
Primary factors productivity	endog.	endog.	endog.	endog.	endog.	−2.5	−2.5
Consumer price index	endog.	endog.	endog.	endog.	endog.	2	2

Notes:
* Endogenous solution
** Because the trend in household numbers is declining, we assume that the annual growth rate of households is 2% from 2000 to 2020.

Table 3.2 The baseline scenario assumptions of TAIGEM©-D

Variables	TAIGEM©-D Baseline Scenario Assumptions
1. Macroeconomy	
1.1 GDP growth rate	GDP growth rates can be either endogenous or exogenous variables.
1.2 Households	Exogenously set at annual growth rate of 2%
1.3 CPI	Annual growth rate is set at 2%
1.4 Industrial structure	Endogenous variable
1.5 Labor (primary factor) demand	Labor is a CES (Constant Elasticity of Substitution) aggregation of the different types of labor force.
2. Energy Structure	
2.1 Liberalization of the oil products market	In the year 2001, the oil products market will be liberalized.
2.2 Nuclear power plant	
2.2.1 2 units of the 4th nuclear power plant	2 units of the 4th nuclear plant will be in operation in the years 2005 and 2006.
2.2.2 1st nuclear power plant retired	In the years 2018 and 2019, the 1st nuclear plant units will be retired at the rate of one unit per year.
2.2.3 2nd nuclear power plant retired	In the years 2021 and 2023, the 2nd nuclear plant units will be retired at the rate of one unit per year.
2.2.4 3rd nuclear power plant retired	In the years 2024 and 2025, the 3rd nuclear plant units will be retired at the rate of one unit per year.
2.2.5 The investment cost of the 4th nuclear power plant	The crowding-out effect generated by the investment cost associated with constructing the 4th nuclear plant is considered.

Table 3.2 (Continued)

Variables	TAIGEM©-D Baseline Scenario Assumptions
2.3 International energy price	Source: Imported oil: EIA/USDOE, International Energy Outlook 2001, Imported steam coal and LNG: IEA/OECD, World Energy Outlook 2000, Imported coking coal and uranium: International Energy Outlook 1998.
2.4 The energy structure	Endogenous variable
2.5 Hydroelectricity	Exogenous, with an annual growth rate of 1.8%.
2.6 Technology bundle	Based on the CRESH function assumption, the elasticity of substitution for hydroelectricity is 0.1, for nuclear power 1.0, for coal 0.1, for oil 0.5, and for LNG 1.5.
3. Adjusting the industrial structure	
3.1 Enhancement of energy efficiency	Endogenous, with the annual energy-saving declining rate set at 0.6%.

in the year 2020. The structure of electricity generation based on our model's predictions indicates that the percentage share of electricity generated from gas goes up over time, while the percentage share of electricity generated from hydro-power goes down. The use of natural gas in power generation rises because the Taiwan Power Company has managed to install more natural gas power plants in consideration of its future responsibility to cope with the potential global warming problem. On the other hand, the construction of hydro-power plants in Taiwan is limited due to the lack of abundant hydro-power resources. As a result, the percentage share of electricity generated from hydro-power falls.

Table 3.5 also demonstrates the other three cases under the assumption that GDP growth rates are exogenous. The high growth rates, middle growth rates and low growth rates are taken from the proposal results of the Energy

Table 3.3 Forecasting the CO_2 emissions baseline with the TAIGEM©-D model

(Unit: Millions of tons of CO_2)

	DGBAS	TAIGEM©-D					
		Exogenous GDP growth rate			Endogenous GDP growth rate		
Year	Real GDP growth rate[1]	Real GDP growth rate[3]	Total emissions	Combustion emissions	Real GDP growth rate	Total emissions	Combustion emissions
1994			167.140	154.626		167.140	154.626
1995	6.42		175.591	162.460	6.38*	175.591	162.460
1996	6.10		185.092	171.398	6.30*	185.092	171.398
1997	6.68		198.530	183.479	7.02*	198.530	183.479
1998	4.57		206.109	190.441	4.57*	206.109	190.441
1999	5.42		221.302	204.350	5.72*	221.302	204.350
2000	5.86		229.81	212.085	5.49*	229.77	212.041
2001	−2.12[2]	2.18	234.920	216.169	−1.46	226.76	208.404
2002	2.23[2]	4.90	244.037	224.693	3.77	232.98	214.233
2003		4.89	253.813	233.905	5.13	242.99	223.657
2004		4.69	263.634	243.232	4.61	252.24	232.445
2005		4.45	268.842	248.023	4.67	258.00	237.765
2006		4.43	276.005	254.744	4.19	264.40	243.769
2007		4.43	286.620	264.840	3.79	272.86	251.833
2008		4.36	297.639	275.336	3.63	281.33	259.923
2009		4.19	308.744	285.928	3.49	289.82	268.034
2010		4.02	319.881	296.562	3.36	298.33	276.166
2011		3.94	331.182	307.366	3.22	306.67	284.155
2012		3.89	342.842	318.521	3.12	315.02	292.154
2013		3.84	354.871	330.041	3.04	323.38	300.168
2014		3.80	367.322	341.980	2.96	331.75	308.203
2015		3.77	380.254	354.397	2.88	340.15	316.265
2016		3.74	393.842	367.462	2.83	348.76	324.531
2017		3.71	407.941	381.044	2.76	357.41	332.849
2018		3.67	422.521	395.118	2.70	366.11	341.231
2019		3.27	440.495	412.683	2.32	375.97	350.808
2020		3.13	454.353	426.155	2.60	386.09	360.649

Notes:

1 The values are the actual values published by DGBAS and the base year is 1996.

2 The value is the forecasted value obtained from the DGBAS's forecasting results.

3 Exogenous GDP growth rates for TAIGEM©-D are estimated values which are obtained from other academic institutions and DGBAS

* The variables are the values of historical simulations from TAIGEM©-D

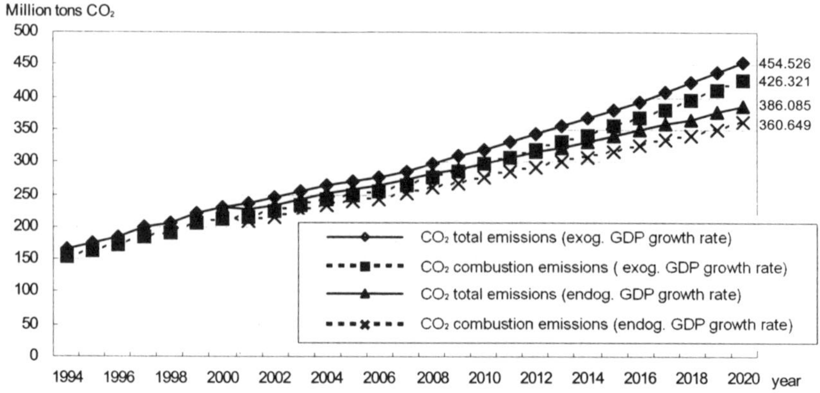

Note: The exogenous GDP growth rates of TAIGEM©-D are estimated values obtained from other academic institutions and DGBAS and which adopt similar assumptions as those stated in Table 3.2.

Figure 3.4 TAIGEM©-D model results: CO_2 emissions baseline

Commission and the Taiwan Research Institute. Three emissions growth projections are specified as 'TAIGEM©-D – high GDP growth rate', 'TAIGEM©-D – middle GDP growth rate', and 'TAIGEM©-D – low GDP growth rate'. The results show that CO_2 emissions in 2020 amount to 443 million tons with the high GDP growth rate, 391 million tons with the medium GDP growth rate, and 346 million tons with the low GDP growth rate. The average growth rates of CO_2 emissions from 2000 to 2020 are 4.41 per cent with the high GDP growth rate, 3.34 per cent with the medium GDP growth rate, 2.40 per cent with the low GDP growth rate, 4.65 per cent with the exogenous GDP growth rate, and 3.24 per cent with the endogenous GDP growth rate.

To evaluate the reliability of the TAIGEM©-D baseline projection, the TAIGEM research group developed a state-space econometric model for purposes of comparison. The state-space projection of Taiwan's CO_2 emissions is listed in the last column of Table 3.5. As the results indicate, the state-space model projects CO_2 emissions in 2015 to be 328 million tons. Unlike TAIGEM©-D, the state-space projection does not take into account the CO_2 emissions generated by the production process in the cement industry.

Therefore, the CO_2 emissions projected by the state-space model tend to underestimate the true emission level by a small amount. However, the growth trend of Taiwan's CO_2 emissions projected by the state-space model is very similar to that generated by TAIGEM©-D. This is an indication of the reliability of the TAIGEM©-D forecasts.

Table 3.4 Comparison of energy structure and electricity structure: 1995–2020

Unit : %

Year	Energy structure					Electricity structure				
	Hydro	Oil	Coal	Gas	Nuclear	Hydro	Oil	Coal	Gas	Nuclear
1995	2.88	52.33	27.35	6.01	11.43	7.38	24.38	34.26	4.61	29.37
1996	2.79	50.38	29.23	5.97	11.64	7.12	19.76	38.67	4.68	29.76
1997	2.80	48.66	31.53	6.40	10.61	7.17	18.88	40.28	6.50	27.18
1998	2.89	47.12	32.58	7.39	10.01	7.32	18.05	39.82	9.61	25.20
1999	2.36	47.72	32.88	6.92	10.12	6.19	18.43	38.62	10.15	26.60
2000	2.26	47.61	33.34	7.17	9.62	5.90	18.50	38.66	11.90	25.04
2001–2005	2.26	47.12	34.13	6.91	9.59	5.81	19.34	38.24	11.91	24.69
2005–2010	2.04	46.26	33.72	6.50	11.47	5.20	18.36	36.99	10.22	29.24
2011–2015	1.94	47.12	34.23	6.58	10.12	4.93	20.12	37.13	12.11	25.71
2016–2020	1.88	48.20	34.56	6.70	8.65	4.71	22.25	36.96	14.39	21.70
Differences										
Year	Energy structure					Electricity structure				
	Hydro	Oil	Coal	Gas	Nuclear	Hydro	Oil	Coal	Gas	Nuclear
2020–1995	−1.00	−4.13	7.21	0.69	−2.78	−2.67	−2.13	2.7	9.78	−7.67

Source: Compiled from TAIGEM©-D baseline forecasting results.

Table 3.5 Comparison of CO_2 emissions baseline forecasting: TAIGEM©-D model and Taiwan-MARKAL model

Unit: Millions of tons of CO_2

Year	Taiwan-MARKAL model*	TAIGEM©-D model: GDP growth rate – High growth	Middle growth*	Low growth	Exogenous	TAIGEM Endogenous GDP growth rate	State–space econometric model
2000	218	212.085	212.085	212.085	212.085	212.041	187.316
2001		222.367	221.111	219.626	216.169	208.404	196.969
2002		230.676	228.074	225.017	224.693	214.233	206.572
2003		239.451	235.401	230.673	233.905	223.657	216.132
2004		248.551	242.942	236.433	243.232	232.445	225.664
2005	250	253.297	246.178	237.97	248.023	237.765	235.162
2006		259.816	251.082	241.208	254.744	243.769	244.627
2007		269.51	258.915	247.121	264.840	251.833	254.054
2008		279.465	266.878	253.056	275.336	259.923	263.454
2009		289.653	274.94	258.981	285.928	268.034	272.826
2010	301	300.041	283.062	264.859	296.562	276.166	282.230
2011		310.495	291.117	270.582	307.366	284.155	291.616
2012		321.089	299.157	276.185	318.521	292.154	300.761
2013		331.809	307.159	281.645	330.041	300.168	310.017
2014		342.647	315.105	286.943	341.980	308.203	319.246
2015	346	353.59	322.979	292.212	354.397	316.265	328.450
2016		364.777	330.897	297.408	367.462	324.531	
2017		376.059	338.728	302.57	381.044	332.849	
2018		387.409	346.457	308.02	395.118	341.231	
2019		400.901	355.849	315.221	412.683	350.808	
2020	404	414.822	365.613	322.63	426.155	360.649	

Note: * GDP growth rates for the TAIGEM©-D and Taiwan-MARKAL models are taken from the exogenous middle GDP growth rates, and both models adopt similar assumptions to those stated in Table 3.2.

Source: Compiled from the projection results of the TAIGEM©-D and Taiwan-MARKAL models.

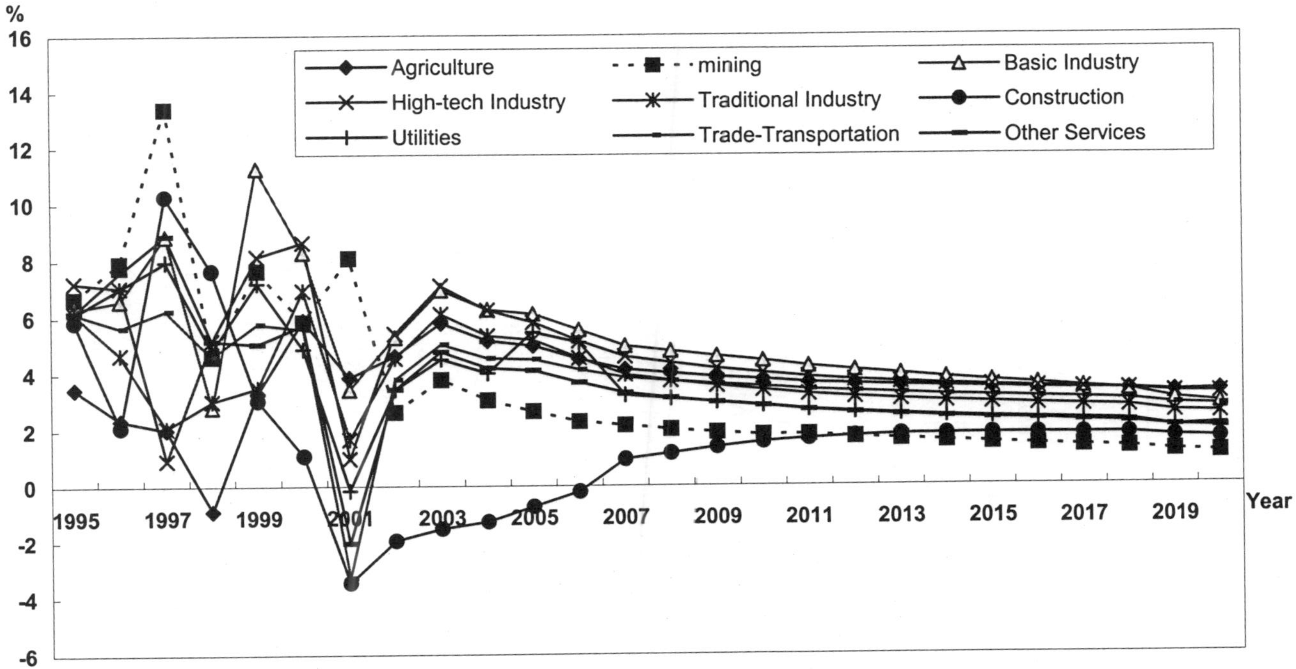

Source: Compiled from the TAIGEM©-D baseline forecasting results.

Figure 3.5 Micro results from the TAIGEM©-D baseline projection – growth rate of industry

Another factor that accounts for the change in the percentage share of power generation from different energy sources is the design of the technology bundle in the TAIGEM©-D model. In power generation, the technology bundle approach enables the least-cost technologies to substitute for the more expensive ones. According to the US Energy Information Administration, the future price of oil may increase. Therefore, the percentage share of oil in power generation may decrease in Taiwan.

Figure 3.5 shows the growth rates of industrial CO_2 emissions in Taiwan taken directly from our TAIGEM©-D baseline projection results using the endogenous GDP growth rates. In this figure, we present the results for nine aggregated sectors: agriculture, mining, basic industry, high-tech industry, traditional industry, construction, utilities, trade-transportation, and other services. The high-tech industry in Taiwan has experienced a fast CO_2 emissions growth in recent years and will maintain a reasonable growth rate in the future. However, the CO_2 emissions growth rates of the mining, traditional industry, and construction sectors decrease significantly over time.

4 COMPARISON OF RESULTS WITH THE TAIWAN-MARKAL MODEL

A key determinant of projected emissions growth is different model structures that are based on different theories. In this section, baseline forecasting results from TAIGEM©-D are compared with those from the Taiwan-MARKAL model, a partial-equilibrium 'bottom-up' model that focuses only on the energy sector and uses exogenous variables for the other sectors. That is, it treats prices exogenously and includes insufficient specifications of the behavior of economic agents outside the energy sector. The Taiwan-MARKAL model has a highly-detailed modeling structure of the energy sector which identifies alternative technologies that can be used in the production of a given output (for example, electricity). Its strength is that it is superior in terms of achieving much greater realism in modeling the substitution options in energy production technology.

By contrast, the TAIGEM©-D model is a general-equilibrium 'top-down' model that embeds Walras's law into the complete economic system. In the model, most of the variables in all the sectors (for example, output prices) are set endogenously and the energy sector is only one of many sectors identified. Interactions among different sectors of the economy through the market mechanism are captured. With the 'technology bundle' approach, we make significant progress in taking MARKAL-like structures into the TAIGEM©-D model.

For baseline projection comparisons, common assumptions are adopted in both the TAIGEM©-D model and the Taiwan-MARKAL model. GDP growth rates up to the year 2020 are set exogenously. A common energy-saving rate is used. Taiwan's oil trade liberalization is expected in the year 2005. Projections from International Energy Outlook 1998 regarding international energy prices are used. Moreover, the amounts of electricity produced with nuclear power are set exogenously.

In regard to the possible structural changes in electricity demand and energy demand, Table 3.4 shows their relative values derived from the TAIGEM©-D forecasting baseline scenario. In the year 2020, the structure of the energy use is biased towards increasing use of coal and gas. By contrast, the usage of oil and nuclear power are decreasing. The major driving force behind this result is that the price of coal is the lowest among all kinds of energy. By adopting the technology bundle approach embedded in TAIGEM©-D, the industry will choose the energy that is relatively cheap to produce. Moreover, according to the projections of the Energy Information Administration (EIA) adopted in the model, the price of oil will increase in the future, and so it will cause the model to reduce the usage of oil.

In terms of the comparison of the structural change in electricity generation in the years 1995 and 2020, the shares of electricity generated from hydro-power, coal and nuclear power plants decline, but the shares of electricity generated from oil and gas increase. Since the baseline projection is based on the assumptions that the amounts of electricity generated from nuclear power remain constant after the year 2000, and that in 2018 the first nuclear plant will be decommissioned, these conditions then cause the trends in nuclear usage quantities in terms of both the electricity structure and the energy structure to decline significantly. Moreover, due to the substitution of different technology bundles, the model tends to increase the usage shares of oil and gas in electricity generation.

Comparisons of fossil fuel combustion emissions projected from the TAIGEM©-D and Taiwan-MARKAL models are presented in Table 3.5 where both models use similar assumptions to those stated in Table 3.2. Of most importance is the adoption of the same exogenous GDP growth rates. The results show that the fossil fuel-related CO_2 emissions projected by TAIGEM©-D for the year 2020 amount to 426 million tons which is higher than the Taiwan-MARKAL model's projection of 404 million tons.

Table 3.6 presents a comparison of model projections in terms of the sectoral emissions structure. The Taiwan economy is further aggregated into five major sectors, that is, the agricultural, industrial, electricity, transportation, and residential–commercial sectors. The results from the two models showing the trend in CO_2 emissions in the electricity sector are quite different. For the year 2020, the Taiwan-MARKAL model projects the electricity sector with a 63.5 per cent share in contrast to the 38.79 per cent share predicted by TAIGEM©-D. It seems that the TAIGEM©-D model is less optimistic in relation to the electricity usage share than the Taiwan-MARKAL model.

The differences in terms of baseline projections between the TAIGEM©-D and Taiwan-MARKAL models can be explained by the following observations:

1. Fossil fuel emissions, process emissions and per capita emissions
The projections of the Taiwan-MARKAL model consider CO_2 emissions from the combustion of fossil fuels only; they do not take into account the emissions from the production process. By contrast, the TAIGEM©-D model is capable of calculating CO_2 emissions from all possible sources.

Because the definitions of the energy demand sectors between the Input–Output Table and the Energy Balance Table are different, TAIGEM©-D reconciles its input–output database with the Energy Balance Table to increase the accuracy of its projections. On the other hand, the Taiwan-MARKAL model solely bases its calculations of CO_2 emissions on the Energy Balance Table. As a result, these two models generate different baseline projections.

2. Sectoral share of CO_2 emissions
As Table 3.6 shows, the CO_2 emissions share of the electricity sector in the TAIGEM©-D projection increases slowly, while the same share in the Taiwan-MARKAL projection increases dramatically. The CO_2 emissions share of the industrial sector generated by the TAIGEM©-D model rises slightly, while the same share in the Taiwan-MARKAL model falls significantly. These results indicate that in the Taiwan-MARKAL projection, the industrial sector shifts a lot of its energy demand from fossil fuels to electricity in its production process. However, the TAIGEM©-D model is more conservative in this regard. Although the future development of the Internet and e-commerce may add to the increased demand for electricity, the predictions of the Taiwan-MARKAL model are apparently more on the optimistic side. According to the Taiwan Energy Commission, energy productivity in Taiwan's industrial sector will decrease because of the expansion of the China Steel Corporation, the operation of the petroleum refining plant of the Formosa Plastics Corporation, and the construction of the Binan Industrial Park for Basic Industries. Therefore, the CO_2 emissions share of the industrial sector may increase in the future. This confirms the predictions of the trend in CO_2 emissions for the industrial sector in the TAIGEM©-D model.

Table 3.6 Sectoral comparison of CO_2 emissions baseline forecasting: TAIGEM©-D model and Taiwan-MARKAL model

Unit: %

Year	Agricultural sector		Electricity sector		Residential–Commercial sector		Transportation sector		Industrial sector	
	TAIGEM©-D	MARKAL	TAIGEM©-D	MARKAL	TAIGEM©-D	MARKAL	TAIGEM©-D	MARKAL	TAIGEM©-D	MARKAL
1996	1.91	n.a.	36.73	n.a.	8.05	n.a.	10.42	n.a.	42.90	n.a.
1997	1.82	n.a.	36.59	n.a.	8.01	n.a.	10.44	n.a.	43.14	n.a.
1998	1.74	0.99	36.67	42.44	8.11	4.71	10.49	15.52	42.99	36.34
1999	1.67	n.a.	36.39	n.a.	7.98	n.a.	10.40	n.a.	43.56	n.a.
2000	1.62	0.9	36.23	45.7	7.94	3.8	10.16	15.6	44.05	34.00
2001–2005	1.54	0.8	36.42	48.0	7.42	3.4	10.04	15.5	44.58	32.3
2006–2010	1.48	0.7	36.69	53.5	7.43	2.9	9.99	14.0	44.41	28.9
2011–2015	1.26	0.6	36.68	58.7	7.37	2.5	9.84	12.5	44.85	25.8
2016–2020	1.08	0.5	38.79	63.5	7.23	2.1	9.62	12.2	43.28	21.7

Note: For the years 1995–99, the sectoral emissions of the TAIGEM©-D model are the results of historical simulations.

Source: Compiled from the projection results of the TAIGEM©-D and Taiwan-MARKAL models based on the middle GDP growth rates assumptions.

The reasons for the above divergences in the models' predictions are complicated. One critical reason is that the projections of the TAIGEM©-D model are based on the historical trend of macroeconomic and microeconomic variables as well as technical change coefficients and consumption taste factors. On the other hand, the Taiwan-MARKAL model envisions more optimistic predictions with regard to innovation and the diffusion of power generation technologies in Taiwan. In fact, the databases for energy technologies and power generation costs used in the Taiwan-MARKAL model are derived directly from the US database without any further modification. The Taiwan-MARKAL model thus tends to overestimate the possibility of Taiwan adopting the advanced US power generation technologies at an early stage. The cost-minimization structure of the Taiwan-MARKAL model introduces many low-cost, high-efficiency energy technologies into the model simulations. However, the cost of power generation in Taiwan is higher than in the US due to the lack of competition and the high cost of acquiring land.

Another important reason is that the Taiwan-MARKAL model attributes all CO_2 emissions from power generation to the electricity sector, while the TAIGEM©-D model allocates CO_2 emissions from power generation to each sector based on its share of electricity consumption. These major differences produce quite different results in terms of the sectoral shares of CO_2 emissions in the Taiwan economy.

Furthermore, energy structure comparisons are provided in Table 3.7. While the Taiwan-MARKAL model is able to incorporate 'new energy' usage into baseline forecasting, the TAIGEM©-D model does not have such a capability due to the lack of 'new energy' data in the input–output database. The Taiwan-MARKAL model projects more usage of oil and nuclear power than does the TAIGEM©-D model. By contrast, more usage of coal, natural gas, and hydro-power is projected by the TAIGEM©-D model.

5 CONCLUDING REMARKS

The major goal of this chapter is to provide a baseline forecast for CO_2 emissions and to compare it with the projections of the Taiwan-MARKAL model for the case of Taiwan. We use the TAIGEM©-D model – a dynamic computable general equilibrium model – to project Taiwan's CO_2 emissions baseline up to the year 2020. According to the TAIGEM©-D model's projections, total CO_2 emissions in Taiwan will increase from 229 million tons in the year 2000 to a significant amount of between 360 and 426 million tons in the year 2020 depending on the annual GDP growth rates adopted by the model. The TAIGEM©-D model has demonstrated its ability to incorporate

Table 3.7 Baseline comparison: energy structure projections: TAIGEM©-D model and Taiwan-MARKAL model

Unit: %

Year	Coal		Oil		Gas		Nuclear		Hydro		New Energy	
	TAIGEM©-D	MARKAL	TAIGEM©-D	MARKAL	TAIGEM©-D	MARKAL	TAIGEM©-D	MARKAL	TAIGEM©-D	MARKAL	TAIGEM©-D	MARKAL
1995	27.35	26.0	52.33	54.0	6.01	10.9	11.43	10.9	2.88	2.8	n.a.	0.6
1996	29.23	n.a.	50.38	n.a.	5.97	n.a.	11.64	n.a.	2.79	n.a.	n.a.	n.a.
1997	31.53	n.a.	48.66	n.a.	6.40	n.a.	10.61	n.a.	2.80	n.a.	n.a.	n.a.
1998	32.58	28.7	47.12	51.0	7.39	9.8	10.01	9.8	2.89	2.8	n.a.	0.7
1999	32.88	n.a.	47.72	n.a.	6.92	n.a.	10.12	n.a.	2.36	n.a.	n.a.	n.a.
2000	33.34	29.7	47.62	50.4	7.17	8.7	9.61	8.7	2.26	2.6	n.a.	0.9
2001–2005	33.98	26.7	47.59	53.4	7.02	10.3	9.25	10.3	2.16	2.1	n.a.	0.9
2006–2010	33.45	30.4	46.58	50.6	6.57	9.1	11.40	9.1	2.01	1.9	n.a.	1.0
2011–2015	33.85	32.1	47.42	49.7	6.66	8.9	10.14	8.9	1.93	1.7	n.a.	1.2
2016–2020	34.12	31.7	48.57	53.3	6.81	6.3	8.61	6.3	1.89	1.4	n.a.	1.2

Source: Compiled from the projection results of the TAIGEM©-D and Taiwan-MARKAL models with the same GDP growth rates, i.e. the middle GDP growth rate.

exogenously-specified GDP growth rates as well as to generate endogenous GDP growth rates by itself in the model simulations. When the adopted GDP growth rates are higher, the projected total CO_2 emissions in Taiwan are also higher. The energy consumption structure of the Taiwan economy is projected by the TAIGEM©-D model to change mildly over the period under study. The consumption shares of coal and natural gas tend to exhibit an increasing trend, while those of oil, hydro-power and nuclear power reflect a decreasing trend. The projected sectoral CO_2 emissions indicate that the high-tech industry in Taiwan has experienced rapid CO_2 emissions growth in recent years and will maintain a reasonable rate of growth in the future. However, the CO_2 emissions growth rate of the mining industry, traditional industry and the construction sector may decrease significantly over time.

The comparison in relation to the baseline forecasting for CO_2 emissions between the TAIGEM©-D model and the Taiwan-MARKAL model indicates a similar pattern of projected results. There are, however, significant differences in the models' projections of CO_2 emissions in the electricity sector and in regard to the structure of energy in the future. The results show that the accuracy of the baseline projections depends on the model being used and the assumptions made. However, the results of the projections for different models, either top-down or bottom-up models, are complementary to each other.

Further improvements in TAIGEM©-D baseline forecasting are underway. They include (1) updating the input–output table to encompass the newly-released 1996 version, (2) the econometric estimation of more elasticities and parameters, (3) extending periods of historical simulation to improve the model's validity, and (4) incorporating other greenhouse gases into the model, for example, methane and nitrous oxide.

NOTES

1. The Taiwan-MARKAL energy model was developed by the Energy and Resources Laboratories of the Industrial Technology Research Institute (ITRI), Taiwan.
2. The CRESH (Constant Ratios of Elasticities of Substitution, Homothetic) function is the generalized form of the CES function, the functional form is $\sum_{i=1}^{n}[X_i/Z]^{h_i}Q_i/h_i = \alpha$, where Z is the output level, the X_i are factor inputs, Q_i, h_i and α are parameters, $0 \neq h_i < 1$, $Q_i > 0$, and $\sum_{i=1}^{n}Q_i = 1$. If $h_i = h$, then the CRESH functional form is reduced to the CES (Constant Elasticity of Substitution) function. The primal distinction between the CRESH function and the CES function is that the substitution elasticities of pairs of factors are equal to some constant in the CES function, but in the CRESH function the substitution elasticity of one pair of factors may be different from that of another pair of factors. For details regarding the CRESH function, see Hanoch (1971).
3. According to the Energy Policy White Papers (see Energy Commission, 1998), the GHG mitigation timetable is set up from year 1995 to year 2020.
4. According to The Trends in Multifactor Productivity ROC published by the DGBAS, from 1979 to 1996, the productivity of primary factors was close to 2.3 per cent on average. For

consistency in terms of the enhancement of energy efficiency with the productivity of primary factors, we set the productivity growth of primary factors at 2.5 per cent in TAIGEM©-D.

REFERENCES

ABARE (1996), *The MEGARABE Model: Interim Documentation*, Canberra, Australia.

Dixon, P.B. and B.R. Parmenter (1996), 'Computable General Equilibrium Modelling for Policy Analysis and Forecasting', in H.M. Amman, D.A. Kendrick and J. Rust (eds), *Handbook of Computational Economics*, **1**, Amsterdam: Elsevier Science.

Dixon, P.B. and M.T. Rimmer (2000), 'Forecasting, Policy, History and Decomposition: The Monash Model of the Australian Economy', available from the Centre of Policy Studies, Monash University, Australia.

Dixon, P.B., M. Malakellis and M.T. Rimmer (1997), 'The Australian Automotive Industry from 1986–87 to 2009–10: Analysis using the MONASH Model', a report to the Industry Commission by the Centre of Policy Studies and IMPACT Project, Monash University.

Dixon, P.B., B.R. Parmenter and M.T. Rimmer (1998), 'MONASH: A Dynamic, Computable General Equilibrium Model of the Australian Economy', available from the Centre of Policy Studies and IMPACT Project, Monash University, Australia.

Dixon, P.B., B.R. Parmenter, J. Sutton and D.P. Vincent (1982), *ORANI: A Multisectoral Model of the Australian Economy*, Amsterdam: North-Holland.

Energy Commisson (1998), *Energy Policy White Papers*, Ministry of Economic Affairs, Executive Yuan, Republic of China, Taiwan.

Hanoch, G. (1971), 'CRESH Production Functions', *Econometrica*, **39**, pp. 695–712.

McDougall, R.A. (1993a), 'Short-Run Effects of a Carbon Tax', General Paper, No. G-100, Centre of Policy Studies, Monash University, Australia.

McDougall, R.A. (1993b), 'Energy Taxes and Greenhouse Gas Emissions in Australia', General Paper, No. G-104, Centre of Policy Studies, Monash University, Australia.

4. Forecasting baseline CO_2 emissions in Japan[1]

Masahiro Kuroda and Koji Nomura

1 INTRODUCTION

At the COP3, held in Kyoto in December 1997, 161 nations participating in the conference agreed to the Kyoto Protocol. The countries agreed to make an effort to reduce emissions of global warming gases from the year 2008 to the year 2010. Although there are doubts if the Protocol will take effect, it is important to examine the likelihood of Japan achieving its target given Japan's predicted situation.[2]

In section 2, we would like to clarify the outlook for environmental protection and energy policy in Japan after COP3, and thus we describe the long-term outlook for Japan's energy supply and demand as well as the present policy. In section 3, we examine the multi-sectoral general equilibrium model for the Japanese economy. Section 4 lays out the details in relation to modeling the energy sector and section 5 lists the assumptions underlying the baseline forecast. The results of the baseline forecast are shown in section 6. Section 7 presents our conclusions.

2 THE ENVIRONMENTAL POLICY AFTER COP3

The Subcommittee for Energy Supply and Demand of the Advisory Committee for Energy prepared an interim report on the 'Long-term Outlook for Energy' in June 1998. This outlook, based firmly on Japan's international pledge regarding the global warming countermeasures of the Kyoto Protocol, describes the energy supply and demand consistent with the targets of environmental protection and economic growth. It assumes an extremely important role by presenting its target and clarifying its stance towards the achievement of a long-term policy.

The Global Warming Prevention Headquarters (hereafter GWPH) formed by the Cabinet in December 1997 reached a consensus regarding the promotion of emission reductions. A consensus was reached in relation to the promotion of emission reductions on such issues as saving energy, countermeasures to reduce emissions of methane and nitrogen dioxide, the reduction of flon gases including HFCs, PFCs and SF6, countermeasures in relation to the source of sink in relation to afforestation, research and development into innovative technology, and the promotion of international cooperation utilizing Joint Implementation and Emission Trading. The GWPH set a target for Japan to reduce greenhouse emissions to 6 per cent below 1990 levels. It was hoped that this reduction would be achieved through the following detailed targets. CO_2, methane and nitrogen dioxide emissions would be reduced to 2.5 per cent below 1990 levels. Through a change in land usage and afforestation activities, carbon sink would be increased by 3.7 per cent. The emissions of HFCs, PFCs and SF6 would be controlled to at least no more than another 2 per cent increase. The last 1.8 per cent was to be obtained through the utilization of Joint Implementation and Emission Trading. The discussions of this Subcommittee for Energy Supply and Demand of the Advisory Committee for Energy were to focus on the long-term outlook for energy supply and demand until the year 2010. Such a scenario was designed to support the comparison of greenhouse gas emissions over this period relative to the 1990 emission levels as agreed in the Kyoto Protocol.

In relation to this long-term energy outlook, by promoting various kinds of energy-saving countermeasures, it is assumed that the final energy supply and demand in the year 2010 can be controlled at the rate of a 0.1 per cent annual increase or 400 million kiloliters of the total amount. This is based on the assumption that the real GNP average growth rate is approximately 2 per cent annually. This assumption is termed 'Business as Usual' (BaU). If the growth of energy can be controlled as assumed above, Japan will save 56 million kiloliters of energy by 2010. These 56 million kiloliters in energy savings are expected to reduce CO_2 emissions from 348 million tons to 287 million tons or approximately 60 million tons by the year 2010.

The possibility of this scenario becoming a reality was avidly discussed by the Energy Supply and Demand Subcommittee. The discussion roughly divided into two viewpoints. One of these was that this target would be nearly impossible to achieve given past observed values of the energy elasticity of GNP. Given the assumed GNP growth rate, energy supply and demand would be much higher. The other viewpoint was that if the countermeasures materialized, they would require a vast reform of the industrial and employment structure. The cost of such a program would impose a heavy burden and would perhaps be too much for the people to carry.

The first problem referred to above concerned whether each of the industry, household and transportation sectors could save the required energy. Various kinds of countermeasures could be adopted, however it would be difficult for the government to achieve the desired result by enforcing regulations. It was hoped that these results could be achieved through the voluntary efforts of industry and individuals. Environmental policies such as the introduction of a carbon tax and an environmental tax were not considered here. It was felt that Japan's economy could not support the additional costs of introducing a new tax system.

On the other hand, the usage of fossil fuels would have to be controlled, because of their contribution to greenhouse gas emissions. Building a new thermal power plant would decrease the dependence on petroleum as part of the total primary energy supply from 51.6 per cent to 47.2 per cent. Assuming that the efficiency of coal combustion will be improved, the share of coal is expected to decrease from 15.4 per cent to 14.9 per cent. On the assumption that new pipelines will be laid and that efforts will be made to privatize natural gas, the share of natural gas is expected to rise from 12.4 per cent to 13.0 per cent.

Regarding non-fossil fuel-based energy, a rapid shift toward atomic energy is expected. According to the 'Supply Scheme of Substitute Energy for Petroleum' promulgated in June 1994, it was expected that, by 2010, 480 billion kWh (70 million kW) of nuclear power plant capacity would be established. However, following accidents at Monju and Fugen and more recently Tokaimura, anti-nuclear sentiment among the people has increased. It is therefore likely to be difficult to add more nuclear power plants in the future. There are 51 commercial nuclear power plants running so far and by the year 2007 nine more will be operable according to the Electric Supply Plan presented in 1998. However, in order to achieve the above projections, the establishment of 10 to 12 more plants is required.

3 THE MULTI-SECTORAL GENERAL EQUILIBRIUM MODEL[3]

Our model was developed to quantitatively analyse the economic interdependence between producers and consumers. The equilibrium price and equilibrium quantity are generated from the market equilibrium conditions in every sector. Producers are divided into 36 industrial sectors, and consumers are divided into six household types classified according to the age of the head of the family, as shown in Table 4.1.[4]

It is assumed that producers take action in two phases in each industrial sector. Given the current situation regarding the capital stock and the number of employees, the short-term supply schedule will be decided by short-term profit maximizing behavior. The supply schedule is dependent on the price of intermediate raw materials including energy and labor services. It has an interdependent relationship in terms of the other industrial sectors through

Table 4.1 Sectoral classification in the multi-sectoral general equilibrium model

Endogenous Sector		Final Demand
1.Agriculture and Forestry	26.Other Manufacturing	37.Household Consumption Expenditures
2.Coal Mining	27.Transportation	38.Government Consumption Expenditures
3.Other Mining	28.Communication	39.Gross Private Fixed Capital Formation
4.Construction	29.Electricity	40.Gross Public Fixed Capital Formation
5.Food	30.Gas Supply	41.Inventory Increase
6.Fiber (Textile)	31.Water Supply	42.Exports
7.Clothing	32.Wholesale and Retail	43.Imports
8.Lumber and Wood Products	33.Finance and Insurance	44.Customs Duties and Import Duties
9.Furniture and Fixtures	34.Real Estate	
10.Paper and Pulp	35.Other Services	Value-added
11.Publishing and Printing	36.Public Services	57.Compensation of Employees
12.Chemicals		58.Capital Income
13.Petroleum Products		59.Net Indirect Tax
14.Coal Products		
15.Rubber Products		Items of Consumption
16.Leather Products		1.Food/Housing/Furniture/Clothing
17.Ceramics and Quarrying		2.Electricity
18.Steel		3.Gas/Other Lighting and Heating
19.Nonferrous Metals		4.Transportation/Communication
20.Metal Products		5.Other Consumption
21.General Machinery		
22.Electrical Machinery		Non-competitive Import
23.Automobiles		37.Crude Oil
24.Other Transport Machinery		38.Natural Gas
25.Precision Machinery		39.Iron Ore

Scrap Sector	Structure of Power Source	Age Group
41.Used Paper	29-1.Atomic Power	1.15–24 years old
42.LPG	29-2.Coal Thermal Power	2.25–34 years old
43.Coke	29-3.LNG Thermal Power	3.35–44 years old
44.Blast Furnace Gas	29-4.Petroleum Thermal Power	4.45–54 years old
45.Scrap Glass	29-5.Hydro/Geothermal Power	5.55–64 years old
46.Scrap Steel		6.over 65 years old
47.Nonferrous Scrap	29-6.New Energy Power	
48.Steel Ships	29-7.Private Power	

Industry Classification	: $IND = \{1, \ldots, 36\}$
Commodity Classification	: $COM = \{1, \ldots, 36\}$
Non-competitive Import Products Classification	: $NCI = \{37, 38, 39, 49\}$
Scrap/By-Products Classification (Input)	: $SBI = \{41, \ldots, 48\}$
Scrap/By-Products Classification (Occurrence)	: $SBO = \{49, \ldots, 56\}$
Raw Material Classification	: $MTR = \{1, \ldots, 12; 15, \ldots, 28; 31, \ldots, 36\}$
Secondary Energy Classification	: $ENR = \{13, 14, 29, 39\}$
Fuel Classification	: $ENF = \{2, 13, 14, 30, 37, 38, 42, 43, 44\}$
Age-Group of the Head of the Household Classification	: $AGH = \{1, 2, 3, 4, 5, 6\}$
Individual Age-Group Classification	: $AGI = \{1, 2, 3, 4, 5, 6\}$
Items of Expense Classification	: $EXG = \{1, 2, 3, 4, 5\}$
Structure of Power Source Classification	: $EPG = \{$29-1, 2, 3, 4, 5, 6, 7$\}$

transactions of intermediate goods and the labor market. The other type of action taken by the producers is to decide on the technological conditions characterizing the industrial technology at the beginning of each period. It is assumed that, given the long-term outlook, factor-related prices, and the direction of technological progress, firms will take action to minimize long-term costs. The long-term cost function utilizes the trans-log formula with regard to each industrial sector. The respective levels of supply and demand in the future and the factor prices of capital, labor, raw materials and energy are utilized to determine the long-term cost-shares of each productive factor. The inputs of capital (K), labor (L), energy (E), and raw materials (M) will be decided in accordance with the decisions related to the cost-shares of each productive factor. After taking depreciation into account, investments will be targeted based on the difference between the capital stock at the beginning of the period and the optimum capital stock.

If the industry is regulated, or if it is an industry with a relatively clear outlook regarding technology, the flexible decision making required for the choice of technologies as those in our KLEM model will most likely lack the required feasibility. Therefore, it may be possible to exogenously provide possible choices relating to the future technologies. Whatever technological scenario is determined endogenously in the KLEM model as well as exogenously along with the outlook for the technology, the real material input coefficients and labor and capital coefficients are characterized as a new technology with the level of capital stock.

The outlook for electricity supply and demand according to the Ministry of International Trade and Industry shows the future power sources being broken down into nuclear, coal-thermal, LNG-thermal, petroleum-thermal, hydroelectric/geothermal, new energy, and self-owned power generation. Because the model determines the total amount of power generation endogenously, the shares of each type of power are also endogenous.

The labor input level is decided in accordance with the optimum level of capital stock. Six age groups classified according to the age of the head of each household are characterized with regard to the units of labor supplied. The labor supply for each household unit is divided into the head and non-heads of the household. According to our observations, we can confirm the following two facts: the labor supply of the head of the household is not sensitive to the wages in the labor market. By contrast, the labor supplies of the non-heads are sensitive to the wages and income of the head of the household. These observations are consistent with the Douglas–Arisawa Law concerning labor supply behavior within households. We therefore try to formulate a model of labor supply behavior that considers heads and non-heads of households separately. The labor supply in terms of hours is aggregated into five individual age groups that are consistent with the labor supply behavior of both heads and non-heads of households. On the other hand, the labor demand in terms of hours is depicted by the KLEM model in each industrial sector and aggregated into five individual age groups. The equilibrium wage rates and employment levels will be determined by the market equilibrium in relation to the labor supply and demand for workers in each individual age group. It is assumed that short-run wages will be based on the contract decided during the previous period in the labor market.

In the final demand sector, the nominal value of government expenditure along with gross public fixed capital formation and inventory investment is provided exogenously. The amount of exports is determined by the export function measured in terms of different goods. The real exports of each product are determined by the exogenous international price relative to the domestic price, and by the global amount of trade, which is also exogenous. With regard to imports, the import share function is set by the relative prices of domestic and imported goods. Private capital formation is determined by the real investment amount in each industry. The magnitude of investment depends roughly upon the observed capital goods structure by industry. Household consumption expenditure is split into two stages. During the first stage, disposable income is determined by aggregating the labor income and capital income by industry. The disposable income after tax is allocated to consumption and saving along with the utility maximizing behavior. During the second stage, the total amount of consumption expenditure allocated during the first stage is distributed into individual goods and services in line with the utility maximizing behavior.

Finally, given the intermediate input coefficients, the Leontief inverse matrix will guide the domestic demand as both the direct and indirect demand lead to the final demand. Prices and income will be adjusted until supply and demand reach an equilibrium. Finally, the generated value added and the final demand will be consistently determined. Simultaneously, the total saving will be balanced with the total investment in equilibrium. The supply and demand equations in relation to money are also formulated. When the so-called IS–LM equilibrium is reached, the interest rate will be decided endogenously. The various parameters of the multi-sectoral general equilibrium model will be estimated econometrically, mostly by utilizing data covering the period from 1962 to 1992.

4 MODELING THE ENERGY SECTOR AND CO_2 EMISSIONS

This section adds some supplementary explanations that link energy inputs to CO_2 emissions. As shown in Table 4.1, there are 36 endogenous sectors, which have been divided into domestic and imported goods. We have classified non-competitive imports into four commodities, along with eight kinds of scrap and by products.

The model provides ratios of materials used to fuel used for different energy sources in each sector so that CO_2 emissions can be calculated for different fuel types. The use of energy can be derived either endogenously from the model or exogenously from an energy use scenario. In the endogenous case, the selection of energy sources is affected by their relative prices. The selection process in the model is described as involving the minimization of long-term costs. Long-term cost curves reflect technological developments, which means that any technological change that results in energy saving will be reflected in the energy input coefficients.

The private non-industrial sector is further divided into business and household sectors. Energy use in business is mostly relevant to 'other services' in the model. Energy use in households is relevant to 'electricity', 'gas and other heating', or 'gasoline used in transportation and communication' based on utility maximization subject to budget constraints. To sum up, energy demand in the model is determined by rational economic activities in the industrial, business, transportation and household sectors, with given relative prices of production elements, income constraints and technological conditions.

With regard to its primary energy supply, Japan relies on imported crude oil and natural gas. Import prices are given exogenously, and selection is made according to the relative prices of different energy sources. Exchange rates are also exogenous. The composition of power sources (installation capacities of nuclear, coal, LNG, petroleum, hydroelectric, new energy and self-owned power) during the period 1985–2010 is presented for reference purposes. The capacity utilization of each source is also provided exogenously. To satisfy the total electricity demand calculated from economic activities, nuclear power, hydroelectric power, new energy and self-owned electric power are used as base load power, and the remaining demand is covered by coal-, gas- and petroleum-thermal power.

The composition of power sources in terms of installation capacities and capacity utilization gives the supply side. The share of each power source is determined endogenously along with the demand side. Changes in the composition of power sources lead to changes in the energy mix of the energy sector. This consequently affects electricity prices and hence the demand for electricity in the industrial, transportation, business and household sectors. Therefore, the solutions to the model for electricity represent a balanced state of electricity demand and supply.

The various coefficients here are fixed on the 1985 standard value. By looking at the time-trend changes over the period 1960–92, the CO_2 emission amount estimated in our model almost corresponds with the emission amount in the 'Energy/Economy Statistics Summary' reported by the Institute of

Energy Economics at the aggregated level. A comparison of the two is presented in Figure 4.1. Although there was a tendency to overestimate before 1985 and a tendency to underestimate after 1985, there is a discernible trend.

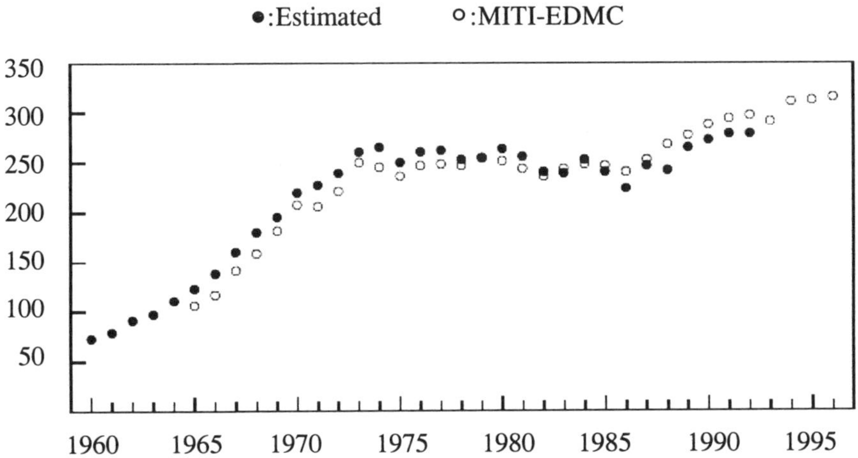

Figure 4.1 The comparison of CO_2 emission amounts

5 ASSUMPTIONS FOR ESTABLISHING THE BASELINE

This section describes the Business as Usual (BaU) scenario. To determine endogenous variables for the BaU case, a scenario that provides exogenous variables through to 2010 is required. The following assumptions are made for the main exogenous variables:

- **Population:** Population increases from 125.6 million in 1995 (reference year) to a peak value of 127.8 million in 2007, and then decreases to 127.6 million in 2010 based on the intermediate estimate by the Institute of Population Problems of the Ministry of Health and Welfare. The population of those aged 15 or over also hits a peak of 111.6 million in 2007, and the number of working-age adults hits a peak of 105.2 million in 2010. The percentage of people who are 65 years of age or over is 20.5 per cent in 1995, which increases to 32.1 per cent in 2010.
- **Number of Households:** The number of households increases from 38.75 million in 1985 to 43.75 million in 1995, and to 51.45 million in 2010. The percentage of single-person households increases from 23.7 per cent in 1995 to 28.6 per cent in 2010.

Table 4.2 Assumptions regarding power sources in the BaU scenario

	Installed capacity (in 10 thousand KW)							Capacity utilization (%)							Labor in Power Industry (in 1,000)	Capital in Power Industry (in trillion yen, 1985 value)
	1	2	3	4	5	6	7	1	2	3	4	5	6	7		
	Nuclear	Coal	LNG	Petroleum	Water	New Energy	Self Power	Nuclear	Coal	LNG	Petroleum	Water	New Energy	Self Power		
1985	2452	1034	2855	5746	3337	0	1507	74.0	64.0	51.2	35.0	28.2	0	51.5	236.1	38.4
1986	2568	1169	2923	5694	3472	0	1547	74.3	55.4	50.9	33.3	26.5	0	55.2	236.0	39.9
1987	2788	1139	3128	5698	3541	0	1609	76.4	63.2	49.6	35.8	24.4	0	55.9	237.5	41.6
1988	2870	1112	3306	5563	3631	0	1689	70.6	65.2	48.8	39.8	28.2	0	58.7	239.2	43.6
1989	2928	1169	3476	5574	3650	0	1826	70.9	65.5	49.6	43.6	28.6	0	58.8	242.0	15.2
1990	3148	1242	3878	5571	3669	0	1966	73.0	66.7	48.7	46.7	28.1	0	57.8	242.8	46.6
1991	3324	1362	3949	5528	3797	0	2039	72.9	66.3	51.4	44.3	29.7	0	58.7	240.3	47.9
1992	3442	1467	4095	5542	3838	0	2130	73.7	67.7	49.0	45.0	25.0	0	57.3	242.3	49.8
1993	3838	1597	4173	5550	3885	0	2249	73.8	68.4	48.2	36.1	29.2	0	56.3	246.4	51.9
1994	4037	1823	4260	5559	4090	0	2321	75.8	66.6	50.5	44.0	20.0	0	56.5	251.9	54.6
1995	4119	2034	4393	5614	4257	0	2405	80.3	66.6	50.5	44.0	23.3	0	57.8	252.3	56.8
1996	4255	2279	4746	5559	4357	10	2440	80.8	62.6	48.7	41.7	22.4	24.7	58.3	258.4	59.2
1997	4492	2524	5100	5504	4347	20	2449	80.8	59.3	47.2	39.4	24.6	23.5	58.3	266.9	61.2
1998	4498	2770	5453	5450	4435	30	2459	77.9	56.6	45.9	37.1	24.6	23.1	58.3	271.1	63.1
1999	4504	3015	5807	5395	4522	40	2468	77.9	54.4	44.7	34.7	24.7	22.9	58.3	275.2	65.0
2000	4510	3260	6160	5340	4610	50	2478	77.9	52.5	43.7	32.2	24.7	22.8	58.3	279.2	66.9
2001	4695	3445	6260	5420	4697	55	2487	77.4	53.3	43.4	31.4	24.5	24.2	58.3	283.9	68.9
2002	4880	3630	6360	5500	4783	60	2497	76.9	54.0	43.0	30.7	24.3	25.3	58.3	288.4	70.9
2003	5065	3815	6460	5580	4870	65	2506	76.5	54.7	42.7	29.9	24.1	26.3	58.3	292.7	72.9
2004	5250	4000	6560	5660	4957	70	2515	76.1	55.3	42.4	29.2	23.9	27.1	58.3	297.0	75.0
2005	5435	4185	6660	5740	5043	75	2525	75.7	55.9	42.1	28.5	23.7	27.9	58.3	301.2	77.0
2006	5620	4370	6760	5820	5130	80	2534	75.3	56.4	41.8	27.8	23.5	28.5	58.3	305.2	79.0
2007	5965	4378	6683	5490	5343	105	2544	76.1	53.4	40.8	31.2	24.5	24.4	58.3	305.4	80.4
2008	6310	4385	6605	5160	5555	130	2553	76.8	50.5	39.7	35.0	25.4	21.9	58.3	305.5	81.7
2009	6655	4393	6528	4830	5768	155	2563	77.4	47.5	38.7	39.4	26.3	20.2	58.3	305.6	83.0
2010	7000	4400	6540	4500	5980	180	2572	77.9	44.6	37.7	44.3	27.1	19.0	58.3	305.7	84.4

Table 4.3 Assumptions regarding exogenous variables in the BaU scenario

		1985	1990	1995	2000	2005	2008	2009	2010	1995 ↕ 2000	2000 ↕ 2005	2005 ↕ 2010	1998 ↕ 2010
	Unit	Volume								Growth rate (%)			
Population	10 thousand	12105	12361	12557	12689	12768	12777	12772	12763	0.21	0.12	−0.01	0.09
No. of households	10 thousand	3875	4084	4375	4666	4925	5060	5103	5145	1.29	1.08	0.88	1.04
Money supply	trillion yen	2952	4291	4609	5246	6224	7018	7304	7602	2.59	3.42	4.00	3.52
Exchange rate	yen/$	238.54	144.80	94.06	122.55	127.56	130.65	131.7	132.76	5.29	0.8	0.8	0.8
Nuclear power capacity	million KW	24.52	31.48	41.19	45.1	54.35	63.1	66.55	70	1.81	3.73	5.06	3.68
Nominal inventory increase	billion yen	2777	2165	3110	3010	3034	2989	2969	2955	−0.65	0.16	−0.52	0.14
Nominal social security benefit	billion yen	34918	48823	64597	82944	106502	123738	130082	136751	5.00	5.00	5.00	5.00
Nominal social security contribution	billion yen	26185	38957	50197	64454	82760	96154	101084	106266	5.00	5.00	5.00	5.00
Nominal governmental consumption expenditure	billion yen	30685	38807	47555	52556	58084	61675	62921	64192	2.00	2.00	2.00	2.00
Nominal governmental capital formation	billion yen	11269	15399	22549	24920	27541	29244	29835	30437	2.00	2.00	2.00	2.00
International trading	10 billion dollars	1936	3466	5147	5605	5892	6072	6133	6194	1.70	1.00	1.00	1.00

- **Government Expenditure:** During the extrapolated period, government consumption expenditure and total fixed public capital formation is assumed to grow by 2 per cent annually, social security benefits and contributions are assumed to grow by 5 per cent annually, and various tax rates remain the same as their current levels.
- **Money Supply:** Nominal money supply is assumed to grow by 3–4 per cent annually.
- **Exchange Rate:** Exchange rates ranging from 238.5 yen/dollar in 1985 to 94.1 yen/dollar in 1995 and to 132.8 yen/dollar in 2010 are projected.
- **Crude Oil Price:** The crude oil price is expected to increase from $18.1/bl in 1995 to $35.1/bl in 2010, which means an average annual growth rate of 4.43 per cent on a dollar basis or 6.73 per cent on a yen basis during 1995–2010.
- **World Trade:** World trade is assumed to increase by 1 per cent annually.

The power source composition scenario in the BaU case is represented by Table 4.2, which is based on the Long-term Energy Supply/Demand Outlook. In the BaU case, changes in capacities between 1996 and 2010 are assumed to increase from 42.55 to 70.00 million KW for nuclear power, 22.79 to 44.00 million KW for coal-thermal power, 47.46 to 64.50 million KW for LNG-thermal power, and 43.57 to 59.80 million KW for hydroelectric power, and so on. The only exception to this pattern of growth is petroleum-thermal power, for which capacity will decrease from 55.59 to 45.00 million KW. Assumptions regarding the labor input and capital stock in the electricity sector are also shown in Table 4.2, estimated according to the operational scheme of each power-generating facility. Table 4.3 presents the assumed schedules of the exogenous variables over the period 1985–2010. The average growth rates are shown in the right-hand part of the table.

6 BASELINE FORECAST RESULTS

Based on the aforementioned assumptions in the BaU scenario, the progress of the main endogenous variables can be derived from the model, as shown in Table 4.4. Real GDP increases from 455 trillion yen in 1990 to 669 trillion yen in 2010. The average annual growth of real GDP is −0.31 per cent between 1995 and 2000, 2.46 per cent between 2000 and 2005, 2.83 per cent between 2005 and 2010, and 1.79 per cent between 1998 and 2010. Regarding the final demand items, the annual average growth rates during

Table 4.4 Simulated BaU results

	Unit	1985	1990	1995	2000	2005	2008	2009	2010	1995 ↕ 2000	2000 ↕ 2005	2005 ↕ 2010	1998 ↕ 2010
		Volume								Growth rate (%)			
Real GDP	billion yen	309716	455016	539683	531514	580948	630308	649603	669205	−0.31	1.78	2.83	1.79
GDP deflator	1985=1	1.026	0.871	0.764	0.914	0.95	0.945	0.941	0.937	3.58	0.79	−0.28	0.89
Interest rate		0.063	0.042	0.049	0.067	0.07	0.068	0.068	0.068	6.04	0.80	−0.33	0.75
CO_2 emission	million t-C	247	298	343	322	331	341	345	349	−1.28	0.53	1.09	0.43
CO_2 emission per capita	t=C	2.0405	2.4138	2.7353	2.5386	2.591	2.672	2.7051	2.7381	−1.49	0.41	1.10	0.34
Fuel consumption	10^{12} Kcal	2075	2581	3033	2852	2940	3038	3075	3111	−1.24	0.61	1.13	0.49
Generated electricity	10 billion KW	60	88	111	116	123	128	130	132	0.85	1.07	1.42	0.97
CO_2 emission per fuel consumption	g/cal	119.01	115.61	113.23	112.96	112.52	112.39	112.36	112.33	−0.05	−0.08	−0.03	−0.06
CO_2 intensity	g/yen	0.798	0.656	0.636	0.606	0.569	0.542	0.532	0.522	−0.98	−1.25	−0.73	−1.36
Energy intensity	kcal/yen	6.701	5.672	5.621	5.365	5.061	4.819	4.733	4.648	−0.93	−1.17	−1.73	−1.30
Average capital service price	1985=1.0	0.985	0.753	0.651	0.87	0.923	0.921	0.918	0.918	5.79	1.18	−1.70	1.30
Average wage rate (age group)	1985=1.0	1	0.977	0.841	1.036	1.139	1.187	1.201	1.213	4.18	1.88	1.27	2.11
Average energy price	1985=1.0	1.025	0.853	0.663	0.729	0.747	0.763	0.769	0.776	1.89	0.50	0.75	0.74
Average raw material price	1985=1.0	1.02	0.842	0.709	0.889	0.956	0.976	0.981	0.987	4.52	1.46	0.64	1.59
Share of nuclear power	%	27.76	24.18	27.02	27.38	30.23	35.27	36.83	36.29				
Share of coal-fired power	%	6.32	8.71	13.67	16.83	18.22	12.69	11.13	11.69				
Share of gas-fired power	%	21.64	20.25	18.19	20.46	20.21	19.62	19.42	19.13				

Table 4.4 (Continued)

		1985	1990	1995	2000	2005	2008	2009	2010	1995 ↕ 2000	2000 ↕ 2005	2005 ↕ 2010	1998 ↕ 2010
	Unit	Volume								Growth rate (%)			
Share of petroleum-fired power	%	18.48	23.61	22.22	15.63	11.98	11.56	11.32	11.89				
Share of hydraulic power	%	13.39	11.29	7.7	8.68	8.64	10.4	10.94	10.78				
Share of new energy	%	0	0	0.02	0.1	0.16	0.22	0.23	0.23				
Share of private power generation	%	12.42	11.96	11.19	10.93	10.56	10.24	10.13	9.99				
Nominal household consumption	billion yen	184685	228938	249008	288123	328238	355326	364777	375077	2.92	2.61	2.67	2.70
Nominal household savings	billion yen	86425	102558	1096963	134938	155268	168464	173033	177798	4.14	2.81	2.71	2.99
Nominal fixed private capital formation	billion yen	80833	105015	52499	72173	84575	89733	90987	92609	6.37	3.17	1.82	3.66
Nominal exports (FOB)	billion yen	45733	45257	68777	86950	102459	116345	121775	127287	4.69	3.28	4.34	3.25
Nominal imports (CIF)	billion yen	35430	37155	30220	40365	49931	57513	60339	63320	5.79	4.25	4.75	4.15
Governmental tax revenue	billion yen	58619	80953	75035	88829	100653	108380	111040	113916	3.38	2.50	2.48	2.61
Capital income tax	billion yen	21048	30445	24090	28077	31904	34212	34990	35886	3.06	2.56	2.35	2.56
Labor income tax	billion yen	18469	26232	26334	31389	35730	38808	39891	40995	3.51	2.59	2.75	2.82
Property tax	billion yen	7684	8859	8843	10766	11913	12473	12637	12827	3.94	2.02	1.48	2.08
Indirect tax	billion yen	10108	13186	13770	15980	17938	19292	19770	20288	2.98	2.31	2.46	2.42
Customs and import duties	billion yen	1310	2230	1998	2616	3169	3595	3752	3919	5.39	3.83	4.25	3.72
Carbon tax	billion yen	0	0	0	0	0	0	0	0	0.00	0.00	0.00	0.00

1998–2010 are 2.7 per cent for nominal household consumption expenditure, 3.66 per cent for fixed private capital formation, 3.25 per cent for exports (FOB prices), and 4.15 per cent for imports (CIF prices).

CO_2 emissions are expected to increase by 349 million t-C (carbon equivalent) in 2010 with an average annual growth rate of 0.43 per cent during 1998–2010. The GDP elasticity of CO_2 during 2000–2010 substantially decreases to about 0.37 because of expanded reliance on nuclear power, compared to a value of 1.26 during 1992–96. The GDP intensity of energy fuels decreases slightly from 5.67 kcal/yen in 1990 to 4.68 kcal/yen in 2010.

In terms of the composition of power sources that incorporate endogenously derived growth in electricity demand, the share of nuclear power increases from 24.18 to 36.29 per cent during 1990–2010. By contrast, the share of petroleum-thermal power decreases from 23.61 to 11.89 per cent over the same period. The share of coal-thermal power increases initially from 8.71 per cent in 1990 to 18.22 per cent in 2005, but decreases to 11.69 per cent in 2010. The share of LNG-thermal power changes slightly from 20.25 per cent in 1990 to 19.13 per cent in 2010.

7 CONCLUSIONS

This chapter presents the current structure of Japan's energy supply and demand as well as policy considerations in response to the Kyoto Protocol. A multi-sectoral general equilibrium model of the Japanese economy is used to construct a baseline forecast to 2010. The forecasted results predict that CO_2 emissions are expected to increase by 0.43 per cent annually during 1998–2010. With an average annual growth rate of GDP of 1.79 per cent, the intensity of energy fuels will fall from 5.67 kcal/yen in 1990 to 4.68 kcal/yen in 2010. The lower intensity of fuel usage can be largely explained by the substitution of nuclear power for petroleum-thermal power. As discussed in the chapter, Japan's CO_2 emissions may increase much faster if Japan hesitates to build the assumed additional nuclear power plants. Nevertheless, the forecasting results depend on the model's assumptions, the expectations regarding the growth rate of the Japanese economy and technological development. A long-term scenario for the baseline forecasting should assume that energy efficiency would be improved in view of industry efforts and technological innovations.

NOTES

1. This chapter was prepared for the 2000 Taipei Conference on Policies for Greenhouse Gas

Reduction and Pollution Control in Asian Pacific, 30 November–1 December, 2000, Taipei. This analysis is a result of a joint research project with Keio University and the Research Center on Global Warming, Development Bank of Japan.
2. See OECD-IEA (1994) and the Government of Japan (1997).
3. See Jorgenson and Wilcoxen (1993), Kuroda et al. (1995), JDB Research Center on Global Warming (1996), and Kuroda and Nomura (1998).
4. See Kuroda et al. (1997).

REFERENCES

The Government of Japan (1997), 'Kyoto Protocol to the United Nations Framework Convention on Climate Change'.

JDB Research Center on Global Warming (1996), 'Symposium on the Environment and Sustainable Development: Roles for Japan with Regard to Global Environmental Issues', The Japan Development Bank.

Jorgenson, D.W. and P.J. Wilcoxen (1993), 'Reducing US Carbon Dioxide Emissions: The Cost of Alternative Instruments', *Journal of Policy Modeling*, **15** (5/6), pp. 491–520.

Kuroda, M. and K. Nomura (1998), 'The Construction of the Multi-Sectoral General Equilibrium Model of the Japanese Economy and the Environmental Protection Policy Simulation: (1) The Environmental Protection Policy Simulation and the Construction of the Multi-Sectoral General Equilibrium Model', KEIO Discussion Paper, No. 15, Keio University Institute for Economic and Industrial Studies.

Kuroda, M., I. Shimpo, K. Nomura and N. Kobayashi (1997), *KEO Database: The Measurement of Production and Inputs of Capital and Labor*, KEIO Monograph Series, No. 8, Keio University Institute for Economic and Industrial Studies.

Kuroda, M., K. Nomura, N. Kobayashi, M. Kuninori, H. Kimiko and H. Tomita (1995), 'Reduction of Carbon Dioxide Emission and Its Distributional Impacts', presented at the JDB Symposium on the Environment and Sustainable Development, Hakone, November, 1995.

OECD-IEA (1994), *The Economics of Climate Change: Proceedings of an OECD-IEA Conference*, Paris: Organization for Economic Cooperation and Development.

5. Analysis of economic and environmental interdependency in East Asian countries

Hikaru Sakuramoto, Kozo Ishida, Masahiko Shimizu, Ayu Washizu and Naoko Takenaka

1 INTRODUCTION

The purpose of this chapter is to analyse the economic and environmental structure of the nine East Asian countries referred to as the EDEN countries: Japan, Korea, China, Malaysia, Singapore, Thailand, Indonesia, the Philippines and Taiwan. To this end, the 1990 East-Asian International Input–Output Tables (EDEN I–O Tables) were compiled in order to analyse the environmental factors and energy efficiency for each East Asian country. The nine EDEN I–O Tables were then linked through each country's foreign trade matrix.

The magnitude of economic and environmental interdependency was calculated by applying the Leontief Open Model. The nine countries were divided into four groups: Japan, the NIEs, ASEAN and China. It was found that: (1) Japan was the largest member of this group and China the smallest. (2) China had the largest CO_2 emissions, followed by Japan and then ASEAN, with the NIEs having the smallest such emissions. (3) An examination of China, ASEAN, the NIEs and Japan indicated the existence of a U-shaped relationship between CO_2 emissions and income per capita, or an inverse 'Environmental Kuznets curve'. (4) China produced the highest CO_2 emissions per unit of final demand with ASEAN second, the NIEs third, and Japan last.

2 ENERGY CONSUMPTION IN EAST ASIAN COUNTRIES

2.1 Structure of Energy Consumption

This section analyses the structure of energy consumption in the nine EDEN countries. The study is based on the EDEN 1990 D table as shown in Figure 5.1. The figure shows aggregate energy consumption in calories.

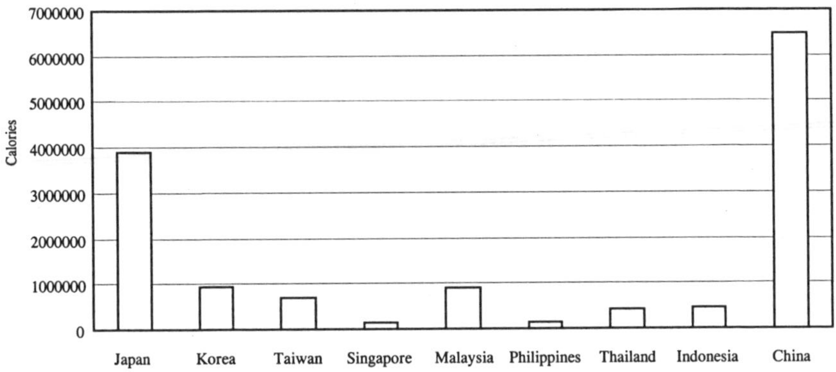

Figure 5.1 Energy consumption in East Asian countries

As a whole, China has the largest energy consumption, followed by Japan, Korea and Malaysia. The countries with the lowest energy consumption in East Asia are the Philippines and Singapore. China and Japan have notably higher energy consumption levels which account for 73.8 per cent of energy consumption in East Asia.

Table 5.1 provides more detailed information on energy consumption across the nine countries. A key distinguishing feature of Japan's energy consumption is that it relies on many energy sources. The four leading sources are diesel oil (15.27 per cent), fuel oil (14.13 per cent), LNG (11.81 per cent), and gasoline (8.925 per cent). Together, these sources account for only 50 per cent of Japan's consumption. Relatively little electricity is used (0.02 per cent) in Japan.

In Korea, fuel oil (19.37 per cent) and diesel oil (15.92 per cent) are the two top sources of energy, and the rates for industrial waste gas (COG) (11.38 per cent) and electricity (10.39 per cent) are comparatively high. Taiwan relies heavily on crude oil (17.68 per cent), coal (15.8 per cent), and electricity (10.36 per cent). Singapore relies heavily on fuel oil (50 per cent) but not on coal (0.39 per cent) or natural gas (0 per cent).

Table 5.1 Energy consumption in EDEN countries

	Japan			Korea		
1	Diesel Oil	595735	15.27%	Fuel Oil	185475	19.37%
2	Fuel Oil	551285	14.13%	Diesel Oil	152469	15.92%
3	LNG	460720	11.81%	Industrial Waste Gas	108958	11.38%
4	Gasoline	348068	8.92%	Total Electricity	99500	10.39%
5	Coal	332460	8.52%	Coal Products	96197	10.05%
6	Coke	307558	7.88%	Coal	78950	8.25%
7	LPG	237604	6.09%	LPG	57637	6.02%
8	Kerosene	234487	6.01%	Coke	50541	5.28%
9	Crude Oil (Inc.Liquid Oil)	187806	4.81%	Kerosene	38658	4.04%
10	Town Gas	158132	4.05%	LNG	28832	3.01%
11	Industrial or General Waste	142177	3.64%	Gasoline	26571	2.78%
12	Petroleum Products	125723	3.22%	Jet Fuel	11560	1.21%
13	Industrial Waste Gas	84166	2.16%	Town Gas	10122	1.06%
14	Jet Fuel	75424	1.93%	Vegetable Fuels	5823	0.61%
15	Petroleum Products	30136	0.77%	Steam and Hot Water Supply	5772	0.60%
16	Natural Gas	18557	0.48%	Naphthas	380	0.04%
17	Naphthas	9078	0.23%	Crude Oil (Inc.Liquid Oil)	50	0.01%
18	Coal Products	1101	0.03%	Natural Gas	0	0.00%
19	Total Electricity	648	0.02%	Petroleum Products	0	0.00%
20	Steam and Hot Water Supply	0	0.00%	Petroleum Products	0	0.00%
21	Vegetable Fuels	0	0.00%	Industrial or General Waste	0	0.00%
22	Natural Energy	0	0.00%	Natural Energy	0	0.00%
	Total	3900866	100.00%	Total	957496	100.00%

Table 5.1 (Continued)

	Taiwan			Singapore		
1	Crude Oil (Inc. Liquid Oil)	210610	29.85%	Fuel Oil	72488	50.60%
2	Fuel Oil	124757	17.68%	Gasoline	21758	15.19%
3	Coal	111447	15.80%	Diesel Oil	16956	11.84%
4	Total Electricity	73113	10.36%	Total Electricity	9476	6.61%
5	Gasoline	43064	6.10%	Crude Oil (Inc. Liquid Oil)	8110	5.66%
6	Diesel Oil	41019	5.81%	Petroleum Products	5080	3.55%
7	Coke	21399	3.03%	Jet Fuel	2956	2.06%
8	Petroleum Products	18462	2.62%	LPG	2439	1.70%
9	LPG	14710	2.09%	Kerosene	2276	1.59%
10	LNG	12904	1.83%	Naphthas	717	0.50%
11	Natural Gas	10870	1.54%	Coal	562	0.39%
12	Jet Fuel	9250	1.31%	Town Gas	302	0.21%
13	Industrial Waste Gas	6855	0.97%	Vegetable Fuels	137	0.10%
14	Naphthas	4549	0.64%	Coke	0	0.00%
15	Kerosene	2487	0.35%	Coal Products	0	0.00%
16	Coal Products	0	0.00%	Natural Gas	0	0.00%
17	Petroleum Products	0	0.00%	LNG	0	0.00%
18	Town Gas	0	0.00%	Petroleum Products	0	0.00%
19	Vegetable Fuels	0	0.00%	Industrial Waste Gas	0	0.00%
20	Industrial or General Waste	0	0.00%	Industrial or General Waste	0	0.00%
21	Steam and Hot Water Supply	0	0.00%	Steam and Hot Water Supply	0	0.00%
22	Natural Energy	0	0.00%	Natural Energy	0	0.00%
	Total	705495	100.00%	Total	143255	100.00%

Table 5.1 *(Continued)*

	Philippines			Thailand		
1	Fuel Oil	42454	31.95%	Diesel Oil	106351	25.60%
2	Diesel Oil	31657	23.82%	Natural Gas	93209	22.44%
3	Total Electricity	19525	14.69%	Fuel Oil	49165	11.84%
4	Gasoline	15884	11.95%	Total Electricity	38177	9.19%
5	Coal	10292	7.75%	Coal	37252	8.97%
6	Kerosene	4767	3.59%	Gasoline	26577	6.40%
7	LPG	4450	3.35%	Vegetable Fuels	18159	4.37%
8	Crude Oil (Inc. Liquid Oil)	2286	1.72%	Industrial or General Waste	13575	3.27%
9	Jet Fuel	1550	1.17%	Jet Fuel	11466	2.76%
10	Natural Energy	12	0.01%	LPG	11096	2.67%
11	Coke	0	0.00%	Crude Oil (Inc. Liquid Oil)	7447	1.79%
12	Coal Products	0	0.00%	LNG	1790	0.43%
13	Natural Gas	0	0.00%	Kerosene	1110	0.27%
14	LNG	0	0.00%	Coke	0	0.00%
15	Petroleum Products	0	0.00%	Coal Products	0	0.00%
16	Naphthas	0	0.00%	Petroleum Products	0	0.00%
17	Petroleum Products	0	0.00%	Naphthas	0	0.00%
18	Town Gas	0	0.00%	Petroleum Products	0	0.00%
19	Vegetable Fuels	0	0.00%	Town Gas	0	0.00%
20	Industrial Waste Gas	0	0.00%	Industrial Waste Gas	0	0.00%
21	Industrial or General Waste	0	0.00%	Steam and Hot Water Supply	0	0.00%
22	Steam and Hot Water Supply	0	0.00%	Natural Energy	0	0.00%
	Total	132877	100.00%	Total	415372	100.00%

Table 5.1 (Continued)

	Indonesia			China		
1	Diesel Oil	104412	23.87%	Coal	4341705	66.86%
2	Kerosene	56181	12.85%	Coke	442242	6.81%
3	Coal	49017	11.21%	Fuel Oil	341021	5.25%
4	Gasoline	40946	9.36%	Steam and Hot Water Supply	249640	3.84%
5	Natural Gas	36158	8.27%	Diesel Oil	247825	3.82%
6	Fuel Oil	34151	7.81%	Industrial Waste Gas	186963	2.88%
7	Total Electricity	32274	7.38%	Gasoline	176042	2.71%
8	LPG	27946	6.39%	Total Electricity	151369	2.33%
9	Vegetable fuels	27309	6.24%	Natural Gas	138816	2.14%
10	Naphthas	19097	4.37%	Crude Oil (Inc. Liquid Oil)	77945	1.20%
11	Town Gas	3236	0.74%	Town Gas	53430	0.82%
12	Industrial or General Waste	2718	0.62%	Kerosene	33064	0.51%
13	Natural Energy	2461	0.56%	Petroleum Products	27699	0.43%
14	Coke	1072	0.25%	LPG	25607	0.39%
15	Coal Products	356	0.08%	Petroleum Products	208	0.00%
16	Crude Oil (Inc. Liquid Oil)	0	0.00%	Coal Products	186	0.00%
17	LNG	0	0.00%	LNG	0	0.00%
18	Jet Fuel	0	0.00%	Jet Fuel	0	0.00%
19	Petroleum Products	0	0.00%	Naphthas	0	0.00%
20	Petroleum Products	0	0.00%	Vegetable Fuels	0	0.00%
21	Industrial Waste Gas	0	0.00%	Industrial or General Waste	0	0.00%
22	Steam and Hot Water Supply	0	0.00%	Natural Energy	0	0.00%
	Total	437333	100.00%	Total	6493762	100.00%

Table 5.1 (Continued)

	Malaysia		
1	Vegetable Fuels	548291	60.86%
2	Natural Gas	294329	32.67%
3	Petroleum Products	33306	3.70%
4	Crude Oil (Inc. Liquid Oil)	11536	1.28%
5	Petroleum Products	5391	0.60%
6	Diesel Oil	4263	0.47%
7	Total Electricity	1715	0.19%
8	Coal	1392	0.15%
9	LPG	642	0.07%
10	Kerosene	110	0.01%
11	Coke	0	0.00%
12	Coal Products	0	0.00%
13	LNG	0	0.00%
14	Gasoline	0	0.00%
15	Jet Fuel	0	0.00%
16	Fuel Oil	0	0.00%
17	Naphthas	0	0.00%
18	Town Gas	0	0.00%
19	Industrial Waste Gas	0	0.00%
20	Industrial or General Waste	0	0.00%
21	Steam and Hot Water Supply	0	0.00%
22	Natural Energy	0	0.00%
	Total	900973	100.00%

Malaysia relies primarily on vegetable fuels (60.86 per cent) and natural gas (32.67 per cent). The Philippines relies on fuel oil (31.95 per cent) and diesel oil (23.82 per cent), and somewhat on electricity (14.69 per cent) and coal (7.75 per cent). Thailand relies on a balance of natural gas (22.44 per cent), diesel oil (25.60 per cent), fuel oil (11.84 per cent), electricity (9.19 per cent) and coal (8.97 per cent). Indonesia relies heavily on diesel oil (23.87 per cent), kerosene (12.85 per cent), coal (11.21 per cent), and electricity (7.38 per cent). China stands out among the EDEN countries in that it depends on coal (66.86 per cent) for most of its energy. Industrial or general waste is used very little in EDEN countries: Japan (3.64 per cent), Thailand (3.27 per cent), and Indonesia (0.62 per cent).

We also examine in detail which industries are tied to each power source. In almost all of the EDEN countries, coal is used primarily for thermal power, especially in Thailand where the rate is 80 per cent. The only country where thermal power is not high-ranking is Singapore. Heavy industries, such as cement, coke and other coal products, iron and steel, and other chemical products, are high energy users. Family consumption is high only in China and Malaysia. Especially in China, families tend to rely on coal for heat. Railway transport is also a high consumer of energy in Indonesia, Malaysia and China.

Which industries rely on natural gas varies significantly across countries. In Japan, natural gas is used largely for gas supply (67.5 per cent). In Taiwan, natural gas is used primarily for family consumption (47.71 per cent). In Indonesia, natural gas is used for thermal power (53.01 per cent). In Thailand, natural gas is split between gas supply (47.91 per cent) and thermal power (44.53 per cent). In China, natural gas is used for crude oil and natural gas (22.2 per cent) and family consumption (16.08 per cent), and the rest goes to manufacturing and electric power.

Diesel oil is primarily used for road transport and secondarily for fishing in Japan, Korea, Taiwan and Thailand. In Malaysia, family industry (46.3 per cent) uses more diesel fuel than road transport. In Singapore, commerce uses 27.59 per cent of the diesel fuel with the rest going to the service industry and road transport.

The structure of energy consumption across countries varies a great deal. Pulp and paper products rely on fuel oil in Korea (71.41 per cent), in Taiwan (55.82 per cent), and in the Philippines (68.5 per cent). Industry in Japan relies on industrial or general waste (41.77 per cent). In Singapore, pulp and paper products rely on electricity (36.65 per cent) and natural gas (46.02 per cent) and, in Malaysia, on vegetable fuels (99.31 per cent). These same companies rely on coal (30.27 per cent) in Thailand and in China (77.79 per cent). Coal accounts for only 10.32 per cent in Japan and 34.13 per cent in Taiwan.

Cement and other non-fabricated metal products rely heavily on coal in all the EDEN countries except Malaysia and Indonesia. In Malaysia, cement relies on town gas (99.30 per cent) and in Thailand it relies on fuel oil (28.99 per cent). In Indonesia, iron relies heavily on coke, as in Japan, Korea, Taiwan and China. In China, the two top industries, coke and coal, use about 70 per cent of total energy.

Iron processes in Singapore, the Philippines and Thailand depend on electricity, whereas in Indonesia, they rely on natural gas (70 per cent). Furthermore, fuel oil is a high-ranked energy in all EDEN countries except Malaysia. Electricity, gas and water rely on liquefied natural gas in Japan, industrial waste gas in Korea, total electricity in Taiwan, crude oil in Singapore, fuel oil in the Philippines, natural gas in both Malaysia and Thailand, and coal in both Indonesia and China. Transport relies on diesel oil in Japan, Korea, the Philippines, Thailand and Indonesia, gasoline in Taiwan, fuel oil in Singapore, and coal in China.

Figure 5.2 shows the structure of five types of energy source for each country. Energy is divided into coal, crude oil, natural gas, electricity and others. Coal is most heavily used in China (73.67 per cent), but hardly at all in Singapore (0.392 per cent). In China, coal is used for electricity, gas and water (31.09 per cent), family consumption (12.96 per cent), cement and other non-fabricated metal products (9.48 per cent), and iron (9.21 per cent). China uses 81.31 per cent of the coal burned in East Asia. Crude oil is heavily used in all EDEN countries except China and Malaysia. Its highest rate of use is in Singapore (90.98 per cent) where it is used in the top three industries: oil and coal products (35.37 per cent), transport (30.12 per cent), and electricity, gas and water (11.81 per cent). In the Philippines, crude oil accounts for 72.2 per cent of energy and the two top industries are transport (37.11 per cent) and electricity, gas and water (29.95 per cent).

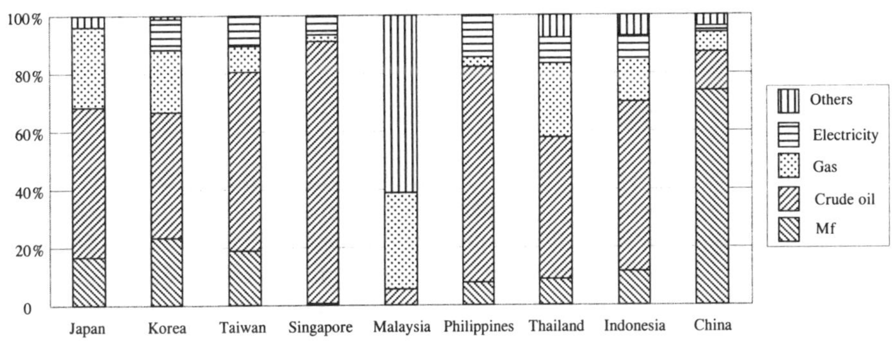

Figure 5.2 Energy consumption composition by country

Electricity is most heavily used in the Philippines (14.69 per cent), Korea (10.39 per cent), Taiwan (10.36 per cent), and Thailand (9.19 per cent), and it is hardly used at all in Japan (0.03 per cent). In the Philippines, the two main industries that use electricity are family consumption (25.53 per cent) and food (11.0 per cent). In Korea, electricity is used for family consumption (20.34 per cent), other services (12.43 per cent), electricity, gas and water (9.59 per cent), and iron (7.25 per cent). In Taiwan, electricity is used primarily for electricity, gas and water (70.8 per cent).

2.2 CO_2 Generation

Figure 5.3 shows CO_2 generation in EDEN countries based on the EDEN 1990 E table. China has the largest CO_2 emissions of the nine EDEN countries. Japan is second, Indonesia is third, Korea is fourth, and Taiwan is fifth. The two countries with the lowest emissions are the Philippines and Singapore and they account for only 2 per cent of China's emissions. China accounts for 53.28 per cent of the region's CO_2 emissions and Japan accounts for another 25.45 per cent.

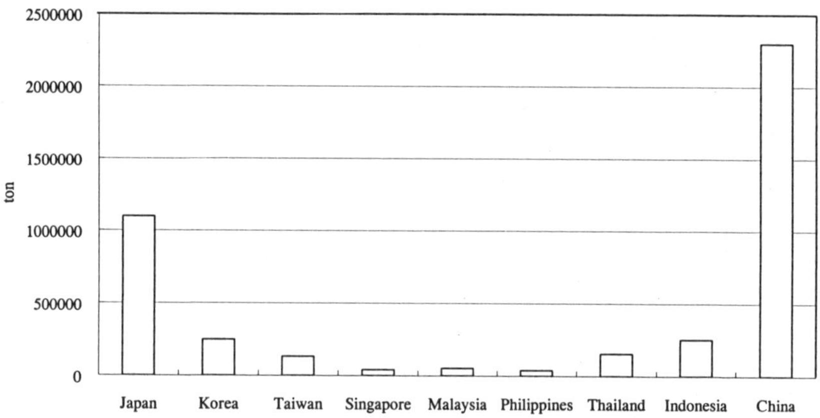

Figure 5.3 CO_2 generation by country

Table 5.2 presents more detailed information about the CO_2 emissions by industry in each country. Thermal power is responsible for a great deal of the CO_2 emissions in Japan (27.70 per cent), Taiwan (29.51 per cent), Thailand (33.43 per cent), Indonesia (50.94 per cent) and China (24.52 per cent). In Japan, family consumption (11.09 per cent), and road transport (10.75 per cent) are also important sources. Korea is the only country where emissions come mainly from family consumption. In Taiwan, thermal power (29.51

Table 5.2 Top ten industries for CO_2 generation

Unit: 1000 Ton

	Japan			Korea		
1	Thermal power	303929	27.701%	Family consumption	53910	21.785%
2	Family consumption	120785	11.009%	Thermal power	41523	16.780%
3	Railway transport	117401	10.700%	Railway transport	22128	8.942%
4	Water, steam and hot water	62252	5.674%	Water transport	11818	4.776%
5	Iron and steel	41609	3.792%	Cement	9807	3.963%
6	Other power	38782	3.535%	Commerce	9018	3.644%
7	Gas supply	34689	3.162%	Gas supply	7744	3.130%
8	Water transport	34348	3.131%	Iron and steel	7303	2.951%
9	Pulp, paper and paper products	31297	2.853%	Coke and other coal products	6136	2.479%
10	Iron and steel products	30349	2.766%	Chemical products	4896	1.978%
	Country total	1097167	100%	Country total	247462	100%

	Philippines			Thailand		
1	Railway transport	11994	31.183%	Thermal power	51513	33.432%
2	Thermal power	9331	24.260%	Gas supply	22560	14.642%
3	Family consumption	2292	5.959%	Railway transport	19814	12.859%
4	Cement	1544	4.015%	Family consumption	7684	4.987%
5	Water transport	1477	3.840%	Cement	6238	4.049%
6	Metal ores mining	1263	3.284%	Water transport	4682	3.039%
7	Other edible crops	1195	3.107%	Other edible crops	4291	2.785%
8	Water transport	988	2.569%	Fishing	3297	2.140%
9	Coke and coal products	797	2.073%	Air transport	3192	2.071%
10	Fishing	783	2.035%	Other power	3063	1.988%
	Country total	38464	100%	Country total	154084	100%

Table 5.2 (Continued)

Unit: 1000 Ton

	Taiwan			Singapore		
1	Thermal power	38805	29.515%	Crude oil	15521	36.942%
2	Railway transport	20604	15.671%	Air transport	6920	16.470%
3	Iron and steel products	13048	9.924%	Thermal power	4903	11.670%
4	Crude oil	8112	6.170%	Water transport	2112	5.026%
5	Chemical products	7797	5.930%	Railway transport	2018	4.802%
6	Water transport	5575	4.240%	Commerce	1786	4.250%
7	Non-ferrous metal products	5427	4.128%	Family consumption	1784	4.247%
8	Iron and steel	4747	3.610%	Other transport	1209	2.877%
9	Family consumption	4119	3.133%	Civic engineering	813	1.936%
10	Fishing	2904	2.209%	Business services	764	1.818%
	Country total	131477	100%	Country total	42016	100%

	Indonesia			China		
1	Thermal power	127918	50.947%	Thermal power	563501	24.527%
2	Cement	36064	14.364%	Family consumption	261091	11.364%
3	Iron and steel	34800	13.860%	Iron and steel	159022	6.922%
4	Gas supply	26019	10.363%	Cement	120476	5.244%
5	Pulp, paper and paper products	5002	1.992%	Chemical products	103369	4.499%
6	Family consumption	3990	1.589%	Non-ferrous metal products	78505	3.417%
7	Non-ferrous metal products	3118	1.242%	Fertilizer	78103	3.400%
8	Civic engineering	2141	0.853%	Iron and steel products	62709	2.730%
9	Railway transport	1936	0.771%	Water, steam, hot water	51318	2.234%
10	Buildings	1641	0.653%	Coke and other coal products	50180	2.184%
	Country total	251080	100%	Country total	2297434	100%

Table 5.2 (Continued)

Unit: 1000 Ton

	Malaysia		
1	Crude oil and natural gas	16618	31.542%
2	Family consumption	12515	23.753%
3	Thermal power	10401	19.741%
4	Railway transport	4254	8.075%
5	Cement	1432	2.719%
6	Commerce	1407	2.671%
7	Fishing	1060	2.012%
8	Other power	740	1.405%
9	Forestry	690	1.310%
10	Buildings	648	1.229%
	Country total	52687	100%

per cent) and road transport (15.67 per cent) are the two top sources of CO_2 emissions.

Singapore's emissions come from petroleum refinery products (36.94 per cent) and air transport (16.47 per cent). In Malaysia, emissions come from crude oil (31.54 per cent), family consumption (23.75 per cent), thermal power (19.74 per cent) and railway transport (8.0 per cent). The Philippines is the only country where road transport (31.18 per cent) is the principal CO_2 source followed by thermal power (24.26 per cent) and family consumption (5.95 per cent). In Thailand, the three top sources of CO_2 are thermal power (33.43 per cent), gas supply (14.36 per cent) and road transport (12.85 per cent). In Indonesia, the principal source is thermal power (50.9 per cent), followed by cement (14.36 per cent), iron and steel (13.86 per cent) and gas supply (10.36 per cent). In China, the two top sources of emissions are thermal power (24.52 per cent) and family consumption (24.52 per cent).

2.3 SO_2 Generation

Figure 5.4 depicts SO_2 generation in EDEN countries based on the EDEN 1990 E table. China is the biggest source of SO_2 generation in the region (66.87 per cent). Japan is the second biggest (14.77 per cent), Korea the third (6.38 per cent), Thailand the fourth (4.78 per cent), and Taiwan the fifth biggest (2.57 per cent). Indonesia (0.31 per cent) and Singapore (0.40 per cent) emit the least SO_2 in East Asia.

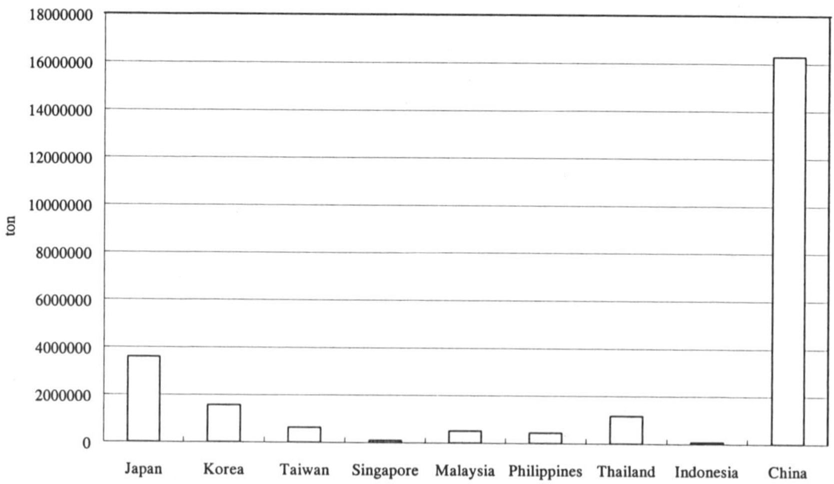

Figure 5.4 SO_2 generation by country

Table 5.3 Top ten industries for SO_2 generation

Unit: Ton

	Japan			Korea		
1	Water, steam, hot water supply	1189494	33.060%	Thermal power	302653	19.450%
2	Thermal power	554588	15.414%	Family consumption	232420	14.936%
3	Iron and steel	472129	13.122%	Water transport	135889	8.733%
4	Pulp, paper and paper products	372948	10.365%	Iron and steel	99757	6.411%
5	Railway transport	178549	4.962%	Chemical products	65477	4.208%
6	Other power	123609	3.435%	Cement	61176	3.931%
7	Water transport	107439	2.986%	Railway transport	55998	3.599%
8	Cement	89120	2.477%	Commerce	49904	3.207%
9	Chemical products	69158	1.922%	Iron and steel products	37669	2.421%
10	Iron products	44382	1.234%	Petroleum refinery products	36072	2.318%
	Total	3597988	100%	Total	1556076	100%

	Philippines			Thailand		
1	Thermal power	179162	40.976%	Thermal power	519843	44.576%
2	Cement	36145	8.267%	Gas supply	130433	11.184%
3	Other edible crops	29509	6.749%	Railway transport	102674	8.804%
4	Railway transport	26407	6.040%	Cement	90024	7.719%
5	Water, steam, hot water supply	24811	5.675%	Water transport	54944	4.711%
6	Metal ores mining	22941	5.247%	Other edible crops	25245	2.165%
7	Water transport	15750	3.602%	Fishing	20952	1.797%
8	Coke and other coal products	10418	2.383%	Non-ferrous metal products	14233	1.220%
9	Pulp, paper and paper products	9659	2.209%	Pulp, paper and paper products	14052	1.205%
10	Commerce	9063	2.073%	Other power	13298	1.140%
	Total	437236	100%	Total	1166204	100%

Table 5.3 (Continued)

Unit: Ton

	Taiwan			Singapore		
1	Thermal power	247527	39.434%	Crude oil	50395	51.509%
2	Chemical products	62108	9.895%	Air transport	9107	9.308%
3	Water transport	54542	8.689%	Thermal power	9046	9.246%
4	Iron and steel products	40714	6.486%	Water transport	6747	6.897%
5	Petroleum refinery products	32170	5.125%	Commerce	4613	4.715%
6	Non-ferrous metal products	27989	4.459%	Other transport	3282	3.355%
7	Pulp, paper and paper products	19186	3.057%	Business services	2086	2.132%
8	Railway transport	16630	2.649%	Railway transport	2083	2.129%
9	Iron and steel	14395	2.293%	Civic engineering	2007	2.051%
10	Spinning	12614	2.010%	Public administration	1399	1.430%
	Total	627697	100%	Total	97838	100%

	Indonesia			China		
1	Thermal power	24787	32.354%	Thermal power	4491244	27.572%
2	Family consumption	6467	8.441%	Family consumption	2093749	12.854%
3	Non-ferrous metal products	4992	6.515%	Cement	989593	6.075%
4	Railway transport	4784	6.245%	Iron and steel	851634	5.228%
5	Civic engineering	4654	6.075%	Chemical products	586350	3.600%
6	Chemical products	4177	5.451%	Non-ferrous metal products	580025	3.561%
7	Buildings	3651	4.766%	Fertilizer	557078	3.420%
8	Other electrical machinery	3628	4.735%	Coke and other coal products	411560	2.527%
9	Cement	3257	4.251%	Water, steam, hot water supply	396778	2.436%
10	Coal products	2346	3.062%	Spinning	367500	2.256%
	Total	76614	100%	Total	16288989	100%

Table 5.3 (Continued)

Unit: Ton

	Malaysia		
1	Family consumption	136255	26.818%
2	Crude oil and natural gas	88699	17.458%
3	Thermal power	71892	14.150%
4	Railway transport	19916	3.920%
5	Commerce	18406	3.623%
6	Cement	14995	2.951%
7	Other electrical machinery	13903	2.736%
8	Other edible crops	11640	2.291%
9	Public administration	10249	2.017%
10	Non-ferrous metal products	9295	1.829%
	Total	508081	100%

Table 5.3 shows which industries are responsible for the SO_2 generation in each country. Water steam, hot water supply and sewage are responsible in Japan for 33.06 per cent of SO_2 emissions. This industry is responsible for substantial emissions in Korea (19.45 per cent), Taiwan (39.43 per cent), the Philippines (40.97 per cent), Thailand (44.57 per cent), and Indonesia (32.35 per cent) as well. In China, thermal power is the main source of SO_2 emissions (25.27 per cent). In Singapore, petroleum refinery products are the worst polluter (51.50 per cent). In Malaysia, family consumption (26.81 per cent) is the biggest SO_2-generating industry. Every country has SO_2 emissions from iron and steel, pulp and paper products, chemical products, cement and coal products. Singapore is the only country that has significant emissions from commerce and service industries.

3 INTERNATIONAL INTERDEPENDENCY

This section describes the interdependency that exists across countries in East Asia. These interdependencies were calculated using the import and export trade matrix of each country in 1990. We combine the EDEN I–O Tables for each East Asian country and these trade matrices. The table also treats the United States and the rest of the world as two regions. The framework of this international EDEN Input–Output table is presented in Table 5.4. This table explicitly examines intermediate demand and CO_2 generation. That is, it calculates the amounts of emissions in other countries that have been caused by the demand for final products in each country.

The equations used to calculate induced calories and induced CO_2/SO_2 generation by each good and service are shown in equations (5.1) to (5.4) below. The Leontief inverse matrix, $(I - A)^{-1}$, is used in our calculation by assuming that goods and services are produced only domestically.

- Induced production

$$X = (I - A)^{-1} F \tag{5.1}$$

- Induced calorie generation

$$Cal = \hat{c}_a (I - A)^{-1} F \tag{5.2}$$

- Induced CO_2 generation

$$\mathrm{CO}_2 = \hat{c} (I - A)^{-1} F \tag{5.3}$$

Table 5.4 Framework of international EDEN Input–Output (1990)

		Intermediate Demand										Final Demand											N	Domestic Production
		Ja	Si	Ta	Ko	Ma	Th	Ph	In	Ch	Us	Ja	Si	Ta	Ko	Ma	Th	Ph	In	Ch	Us	Ex		
Intermediate Input	Ja																							
	Si																							
	Ta																							
	Ko																							
	Ma																							
	Th																							
	Ph																							
	In																							
	Ch																							
	Us																							
	Fi																							
	Ro																							
	Du																							
Value Added																								
Domestic Production																								

Calorie										
CO_2										
SO_2										

Notes:

Ja: Japan	Fi: freight and insurance	In: Indonesia
Ko: Korea	Ph: Philippines	Ro: Rest of the world
Si: Singapore	Us: United States	Du: Duty (excise tax, tariff and etc.)
Ma: Malaysia	Ch: China	Ex: export to the rest of the world
Th: Thailand	Ta: Taiwan	N: statistical discrepancy

- Induced SO_2 generation

$$SO_2 = \hat{s}(I - A)^{-1} F \tag{5.4}$$

X: vector of domestic production;
I: identity matrix;
A: input coefficient matrix;
F: vector of final demand.
$\hat{c}_a$, $\hat{c}$, $\hat{s}$: diagonal matrices of calorie coefficients; and
CO_2 and SO_2 are generation coefficients per unit of production.

Equation (5.1) is used to calculate induced production by final demand, both direct and indirect. Equations (5.2), (5.3) and (5.4) are used to calculate induced energy, induced CO_2 and induced SO_2 generation, both direct and indirect, based on the increase in final demand with the use of each coefficient of generation per unit of production. The estimations were performed for the nine countries mentioned above.

Economic development in the Asian region has followed the flying geese pattern. Developing countries catch up with developed countries in the short term by obtaining technologies and experience from their more developed neighbors. The NIEs have sought to catch up with Japan, which has accomplished its economic development already. The ASEAN (Association of South-East Asian Nations) countries have been catching up with the NIEs, and China is now following the ASEAN countries. Therefore, in the following analysis we divide the nine countries into four groups (Japan, NIEs, ASEAN and China), each of which is at a different stage of economic growth.

3.1 Structure of Energy Consumption

The structure of energy consumption in the four groups is reviewed based on the EDEN 1990 D Table. Figure 5.5 shows the five types of energy consumed in each group. China consumes the most energy followed by Japan, the ASEAN countries and the NIEs. Half of the energy in Japan and the NIEs comes from crude oil. Electricity is comparatively high in the NIEs (10.08 per cent), but very low in Japan (0.02 per cent). In the ASEAN countries, vegetable fuels account for 32.47 per cent of the energy, and coal is comparatively low (5.27 per cent). Coal is very important in China.

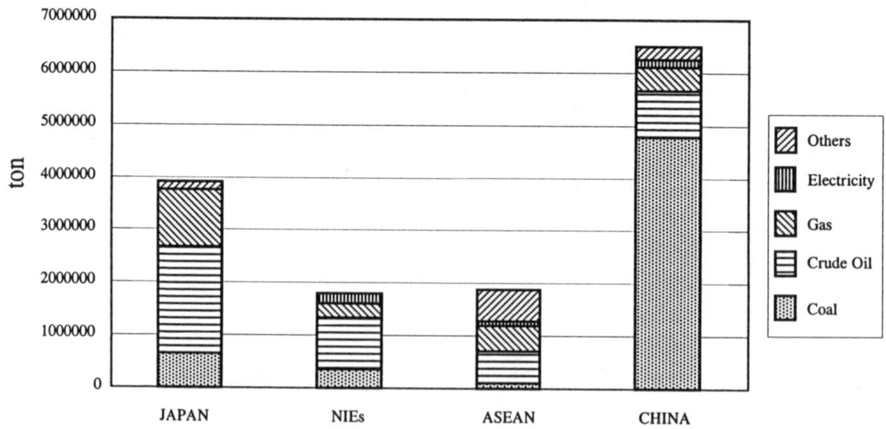

Figure 5.5 Energy consumption structure by group

3.2 Induced Production Based on Final Demand

Figure 5.6 shows how much the final demand of each group is met by domestic production versus production in other groups. The figure includes the international interdependency of the four groups. For example, if the final demand of Japan were to increase, both domestic production and imports would increase.

Figure 5.6 shows that the final demand of Japan results in a large amount of production in East Asia. The quantity of production based on the final demand of the ASEAN countries is the lowest among the four groups. The rate of production based on the increase in final demand is 98.19 per cent in Japan, 90.48 per cent in the NIEs, 85.38 per cent in ASEAN, and 96.90 per cent in China. That is to say, it is very high in Japan, but very low in ASEAN. As to the rate of production in relation to the other groups, ASEAN has the highest rate of the four groups.

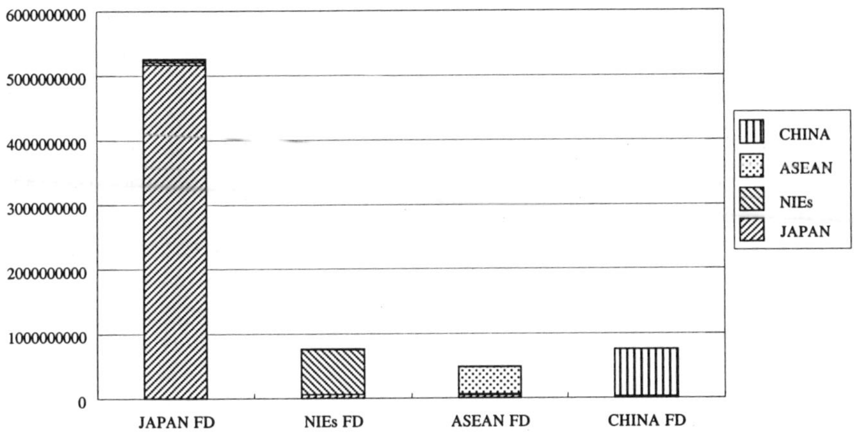

Figure 5.6 Induced production and final demand by group

Figure 5.7 shows how production would change if final demand were to increase by $1 billion in each group. Japan's production is found to be the most sensitive to changes in final demand in the other countries. A $1 billion increase in demand raises Japanese exports by $78.5 million in relation to the NIEs, by $95 million for the ASEAN countries and by $19.8 million for China. The only other large interaction occurs between the NIEs' demand and ASEAN production. The remaining links across countries are relatively small.

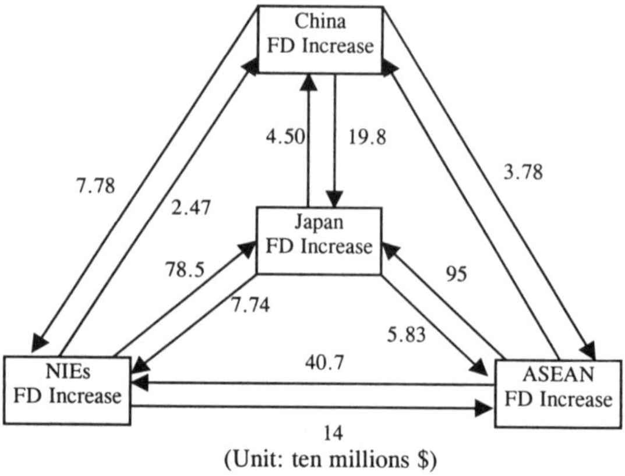

Figure 5.7 Induced production by $1 billion increase in final demand between groups

Table 5.5 shows how industries are affected by the increases in final demand. In all cases, 'other services' is the most sensitive industry, and commerce, transport, construction, civil engineering and food are ranked among the top ten industries except for construction in China. The top ten industries are more or less the same in Japan, the NIEs and ASEAN. Dairy farming, edible crops, spinning and weaving are highly-ranked industries only in China.

An increase in Japanese demand results in an increase in tertiary industries in Japan, in heavy industries such as electrical machinery, iron and chemical products in NIEs, in forestry, timber, wooden products, petroleum refinery and coal products in ASEAN, and in light industries such as spinning and weaving in China.

An increase in demand in NIEs causes an increase in heavy industries such as machinery or iron and services in Japan, in tertiary industries in the NIEs, in electrical machinery, forestry and timber and wooden products in ASEAN, and in edible crops, spinning and weaving in China. An increase in ASEAN demand causes an increase in services, commerce and heavy industries such as general and electrical machinery, iron, motor vehicles and chemical products in both Japan and the NIEs, in tertiary industries such as services, commerce and transport in ASEAN, and in iron, spinning, weaving and edible crops in China.

Finally, an increase in final demand in China causes an increase in the secondary and tertiary industries such as iron, machinery, commerce, financial and insurance services in both Japan and the NIEs, in timber, wooden products, chemical products, edible crops and commerce in

Table 5.5 Structure of energy consumption by industry in each group, 1990

		Pulp, paper & paper products		Cement, non-fabricated metal products		Iron		Electricity, gas & water		Transport		Total of five	
JAPAN	Coal	10133	10.72%	67294	63.54%	285298	101.05%	217198	15.18%	40	0.01%	579963	22.89%
	Crude Oil	41334	43.72%	31609	29.84%	20252	7.17%	505182	35.30%	568851	91.72%	1167228	46.06%
	Gas	3560	3.77%	6553	6.19%	-23279	-8.25%	606437	42.38%	51275	8.27%	644545	25.44%
	Electricity	28	0.03%	14	0.01%	57	0.02%	19	0.00%	27	0.00%	145	0.01%
	Others	39493	41.77%	444	0.42%	0	0.00%	102108	7.14%	0	0.00%	142045	5.61%
	Total	94549	100.00%	105914	100.00%	282328	100.00%	1430943	100.00%	620193	100.00%	2533927	100.00%
NIEs	Coal	2766	15.84%	42627	63.56%	65510	48.10%	95890	26.31%	433	0.16%	207227	24.34%
	Crude Oil	11943	68.39%	17719	26.42%	17990	13.21%	117636	32.27%	237793	89.35%	403081	47.34%
	Gas	49	0.28%	2263	3.37%	43269	31.77%	89577	24.57%	25717	9.66%	160875	18.90%
	Electricity	2685	15.38%	4441	6.62%	9354	6.87%	61430	16.85%	2189	0.82%	80098	9.41%
	Others	19	0.11%	18	0.03%	83	0.06%	0	0.00%	5	0.00%	125	0.01%
	Total	17462	100.00%	67068	100.00%	136206	100.00%	364533	100.00%	266137	100.00%	851407	100.00%
ASEAN	Coal	738	2.84%	21987	6.93%	70	0.22%	67620	19.23%	16	0.01%	90430	9.79%
	Crude Oil	2796	10.77%	14281	4.50%	3711	11.72%	89398	25.43%	191462	97.27%	301649	32.67%
	Gas	871	3.35%	7834	2.47%	6048	19.09%	183086	52.07%	4153	2.11%	201991	21.88%
	Electricity	1343	5.17%	3970	1.25%	3159	9.97%	9023	2.57%	1200	0.61%	18695	2.02%
	Others	20217	77.86%	269069	84.84%	18692	59.00%	2473	0.70%	13	0.01%	310465	33.63%
	Total	25965	100.00%	317141	100.00%	31680	100.00%	351600	100.00%	196844	100.00%	923230	100.00%
CHINA	Coal	72539	78.22%	453900	89.65%	440760	66.37%	1487786	86.62%	91535	36.95%	2546519	78.87%
	Crude Oil	4792	5.17%	35960	7.10%	43793	6.59%	162618	9.47%	153496	61.96%	400658	12.41%
	Gas	616	0.66%	5945	1.17%	135596	20.42%	43363	2.52%	600	0.24%	186121	5.76%
	Electricity	2928	3.16%	8379	1.66%	13778	2.08%	17818	1.04%	1985	0.80%	44887	1.39%
	Others	11861	12.79%	2145	0.42%	30199	4.55%	6122	0.36%	138	0.06%	50466	1.56%
	Total	92736	100.00%	506329	100.00%	664126	100.00%	1717707	100.00%	247753	100.00%	3228650	100.00%
TOTAL	Coal	86176	37.35%	585807	58.79%	791638	71.04%	1868494	48.35%	92024	6.91%	3424139	45.43%
	Crude Oil	60865	26.38%	99569	9.99%	85746	7.70%	874834	22.64%	1151602	86.53%	2272616	30.15%
	Gas	5096	2.21%	22595	2.27%	161634	14.51%	922462	23.87%	81745	6.14%	1193532	15.84%
	Electricity	6984	3.03%	16804	1.69%	26348	2.36%	88289	2.28%	5400	0.41%	143825	1.91%
	Others	71590	31.03%	271676	27.26%	48975	4.40%	110703	2.86%	156	0.01%	503101	6.67%
	Total	230712	100.00%	996452	100.00%	1114340	100.00%	3864783	100.00%	1330927	100.00%	7537214	100.00%

ASEAN, and in services, dairy farming, machinery, spinning and weaving in China.

3.3 Induced CO_2 Generation

This section deals with the energy and environmental problems associated with production but not consumption, based on final demand. We calculate induced CO_2 generation using equation (5.3). Figure 5.8 shows the total quantity of induced CO_2 generation by the four groups.

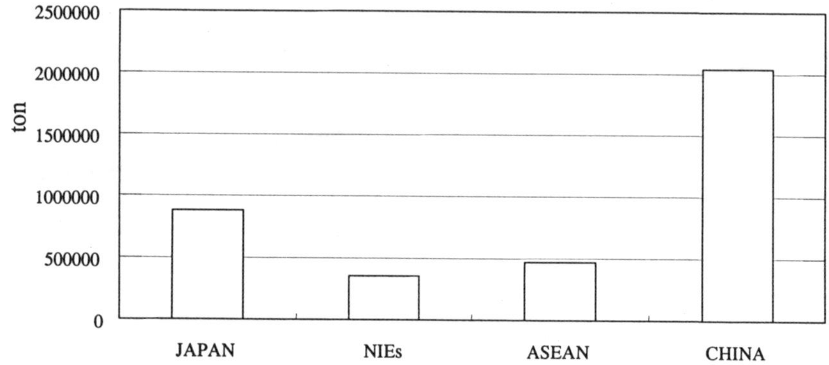

Figure 5.8 Induced CO_2 generation by group

China's emissions are 2.23 times those of Japan and 5.75 times those of NIEs. China produces 54.48 per cent of total induced CO_2 generation in East Asia. Japan produces 23.4 per cent of induced CO_2 generation, the NIEs produce 9.47 per cent, and ASEAN produces 12.54 per cent of regional emissions.

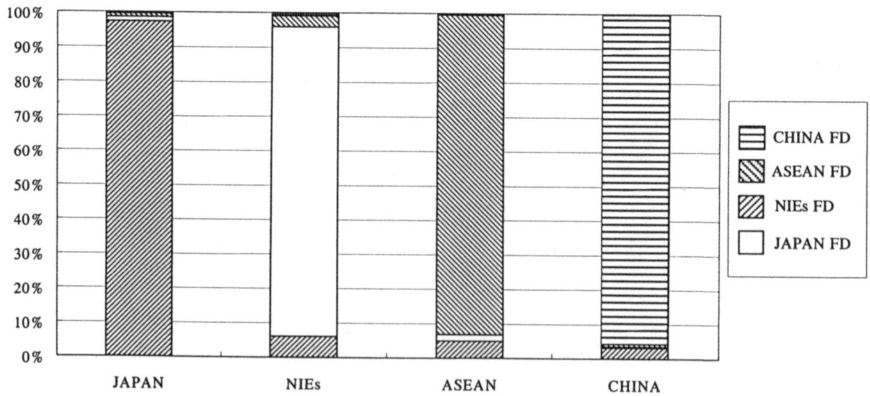

Figure 5.9 Induced CO_2 generation by sources

Figure 5.9 is more detailed than Figure 5.8 and shows the structure of induced CO_2 generation together with the production process in each group. According to Figure 5.9, CO_2 emissions are largely due to production for domestic consumption. The interrelationships across the four groups are actually quite small. The largest effect in Figure 5.9 is that for the NIE's where 10 per cent of production is for consumption in other countries.

Figure 5.10 shows induced CO_2 generation per head of population in 1990 for each group. Japan has the highest per capita emissions of CO_2 followed closely by the NIEs. Both ASEAN and China have surprisingly low emissions on a per capita basis.

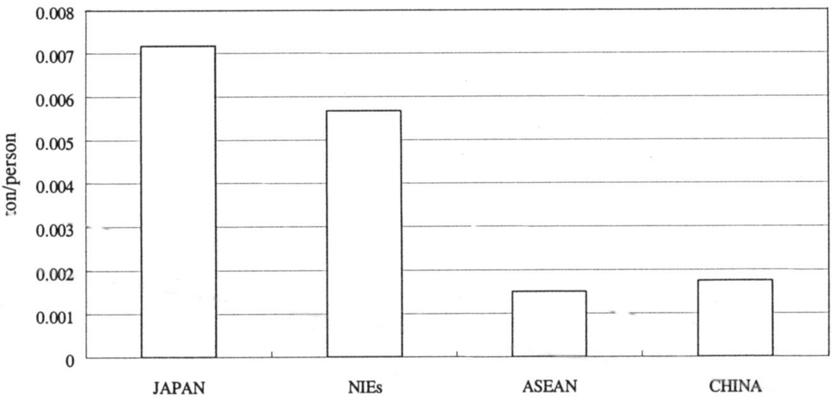

Figure 5.10 Per capita induced CO_2 generation by group

Figure 5.11 shows induced CO_2 generation per unit of final demand. According to Figure 5.11, it is ASEAN that has the greatest induced CO_2 generation per unit of final demand, followed by the NIEs, and then China and Japan. China, which has the lowest income among the four groups, is the country with the lowest induced CO_2 generation in terms of final demand, because its demand structure does not depend on goods and services that induce high quantities of CO_2. However, ASEAN, which has higher incomes and is more industrialized than China, has the most induced CO_2 generation among the four groups, because its structure of demand depends on goods and services that induce a large amount of CO_2 and lack of corresponding environmental measures.

The NIEs have higher incomes but smaller CO_2 generation than the ASEAN countries, because they depend on an energy-saving environment as a result of the progress of industrialization and services. Finally, Japan, which has the highest income among the four groups, is the smallest group in terms of amount of CO_2 generated because of the progress in the services industry and technology and because the system of environmental regulations is already arranged to some extent.

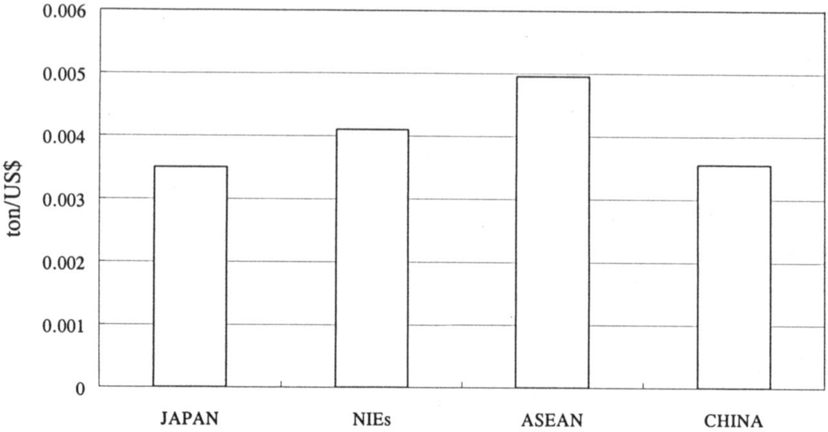

Figure 5.11 Induced CO_2 generation per final demand

According to the White Paper on the Environment (1997), the path of 'China → ASEAN → the NIEs → Japan' in terms of induced CO_2 per final demand generation is a reversed U-shaped distribution, which might be called the 'Environmental Kuznets curve'. If the income in China were to increase steadily, it would experience a similar pattern to ASEAN which has the largest induced CO_2 generation, followed by the NIEs, and finally it would experience a similar pattern to Japan in the future. This figure also suggests that we have to drag down the 'Environmental Kuznets curve' to create an ecology-friendly and energy-saving environment in East Asia, by encouraging a change in the industrial structure, and by establishing environmental safeguards in each country and enterprise.

Figure 5.12 suggests that Japan is the smallest group in terms of the quantity of induced CO_2 generation per unit of final demand, but it has a structure of demand which induces production of CO_2 goods from the other groups, especially from China. Similarly, the NIEs depend on goods that induce more CO_2 emissions to ASEAN and China than to Japan by comparison. That is to say, the NIEs have a structure of demand that depends on energy-saving goods going to Japan, but not on those going to ASEAN and China.

In addition, ASEAN, which is considered the area least dependent on energy-saving goods as mentioned above, has a demand structure that depends on energy-saving goods going to Japan and the NIEs, but not on energy-saving goods going to China. As for China, it has a demand structure that depends not on energy-saving goods going to China itself, but on energy-saving goods going to the other three groups. So there is a common feature that Japan is relatively dependent upon energy-saving goods from the other three groups. By contrast, China is not dependent on energy-saving goods from the other three groups. When viewed from another perspective, only for Chinese goods that are made in China and ASEAN goods that are made in ASEAN can we find the 'Kuznets curve' mentioned above, and so we understand how important it is to change the industrial structure and bring in new technologies to protect the environment.

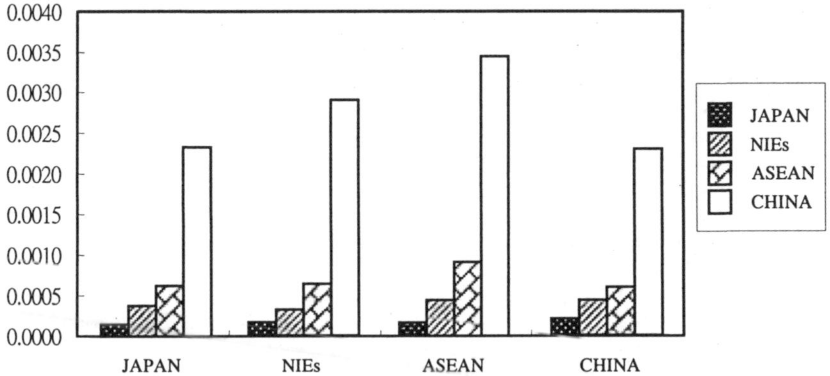

Figure 5.12 Induced CO_2 generation per unit of final demand

4 FINAL REMARKS

The problem of global warming is the ultimate environmental problem facing human beings because it affects the stability of the global climate. During the period 1980–95, total CO_2 emissions for the world increased 1.2 times. At the same time, the rate for East Asia increased 1.9 times, somewhat faster than that for the world as a whole. The total CO_2 emissions in East Asia apart from Taiwan accounted for 14.87 per cent of world emissions in 1980, 16.9 per cent in 1985, 19.37 per cent in 1990, and 24.04 per cent in 1995. So the rate has grown year by year.

It is difficult to establish a consensus regarding environmental problems among countries, because each country has a different background and sense of values. We have to think about how each country can establish good relationships with other countries, and it may not be clear who is the villain and who the victim. The circumstances in East Asia have been described in this study. Based on the international input–output analysis, we can understand the interdependence among countries in this region in order to clarify to what extent each country will be affected by other countries' economic activities and technology level. As can be seen, an interdisciplinary working framework and further international cooperation to engage in economic and environmental analysis is necessary to foster sustainable economic growth for all the Asian economies.

REFERENCES

Ishida, Sakuramoto, Masahiko Shimizu and Takenaka Wasizu (2000), 'Analysis for Economic and Environmental Interdependency in East Asian countries', KEO Discussion paper No. 78, Keio Economic observatory.

Kiji, Sasaki and Kimu Wasizu (2000), 'EDEN Data Base', KEO Discussion paper, No. 79, Keio Economic observatory.

The Ministry of Environment (1997), *White paper on the environment*.

Miyazawa, K. (eds) (1996), *Sangyorenkanbunseki-nyumon*, Nihon-keizai.

Ono, T. and Y. Katou (1998), 'The change of interdependency in East Asia and JAPAN, USA, and EUROPE', *Innovation & I-O technique*, **8** (3), PPAIOS, 1998.

Ono, K. and K. Sakurai (1997), 'Developed economy of East Asia', *Yuhikaku ARMA*, 1997.

Ueda, K. (1998), 'Kankyokeizaigaku', Iwanami.

The United Nations (1995), *Statistical Yearbook*, United Nations Publication, New York, USA.

PART II

Abatement Costs

6. Effect of energy tax on CO_2 emissions and economic development in Taiwan, 1999–2020

Chi-Yuan Liang and Dale W. Jorgenson

1 INTRODUCTION

In response to the new developments subsequent to the Kyoto Protocol of 11 December, 1997, the government of Taiwan hosted a nationwide Energy Conference in May 1998. Two important recommendations that emerged among others at this conference were that (1) the targeted reduction in CO_2 should be 24 to 48 per cent lower than the BaU (business as usual) projection by 2020; (2) economic incentives such as an energy tax, a carbon tax and emissions trading should be considered and carefully examined.

Since July 1995, Taiwan has implemented an anti-air-pollution fee (AAPF) to reflect the social cost of air pollution and to reduce air pollution. The tax revenue from the AAPF is used for anti-air-pollution purposes only. The implementation of an energy tax or a carbon tax might be relatively more acceptable from the point of view of Taiwan's taxpayers as compared with taxpayers in other countries, because the logic behind implementing an energy or carbon tax, which should reflect the social cost of CO_2 emissions, is analogous to that behind the existing AAPF.

Furthermore, Taiwan's energy prices are relatively lower than those of its neighboring trading partners. For instance, Taiwan's gasoline and diesel prices were NT$19.2 and NT$34.29 per liter, respectively, in September 2000. They were 44 per cent and 27.5 per cent, respectively, lower than the corresponding averages for Japan, Korea, Hong Kong and Singapore. This is due to the fact that Taiwan's tax rate on oil products is the lowest among the five economies mentioned above.

It is pertinent for Taiwan to raise its energy prices via higher energy-related taxes in order to conserve energy consumption and to reduce the social costs associated with energy consumption, which arise because of air pollution, CO_2 emissions, traffic congestion and instability of energy supply. However, whether the implementation of energy-related taxes is plausible or not depends on a precise evaluation of the effect of the taxes on energy conservation and the economy.

Since a high energy tax might have a significant impact on the economy, a step-by-step or 'progressive' approach should be examined as an alternative. Therefore, once it is decided to implement an energy tax, the next question will be that of determining the best approach, that is a 'one-step' approach or a 'progressive' approach. The selection will depend on a comparison of the energy tax effects of a 'one-step' approach and a 'progressive' approach on CO_2 emissions and the economy.

The purpose of this chapter is therefore to evaluate and compare the effect of an energy tax on the price level, output growth and CO_2 emissions by sector and for the economy as a whole by applying the 'one-step' approach as well as the 'progressive' approach during the 1999–2020 period.[1] Policy recommendations are drawn from the findings.

The chapter consists of the following four sections: (1) introduction; (2) the theoretical model; (3) the simulation methodology and procedure; (4) simulation results; and (5) conclusion.

2 THE THEORETICAL MODEL – DYNAMIC GENERALIZED EQUILIBRIUM MODEL OF TAIWAN

The dynamic generalized equilibrium model of Taiwan (DGEMT) consists of the following four sub-models: (1) the producer's model; (2) the consumer's model; (3) the DGBAS's macroeconomic model; and (4) ITRI's MARKAL engineering energy model.

2.1 The Producer's Model

The producer's model breaks down the Taiwan economy into 29 sectors, namely, eight main sectors (agriculture, mining, manufacturing, construction, public utilities, transportation, services and industry (mining, manufacturing, construction and public utilities)), 17 manufacturing sectors (food, beverages and tobacco, textiles, clothes and wearing apparel, leather and leather products, wood and bamboo products, furniture products, paper and printing, chemicals and plastics, rubber products, non-metallic minerals, basic metals, metal products, machinery and equipment, electrical machinery and electronics, transportation equipment and miscellaneous), and four energy sectors (coal mining, oil refining, natural gas and electricity).

We assume that the sectoral cost function is of the translog form with homothetic weak separability of energy and material inputs. The model actually consists of four sub-models (for each sector): an aggregate sub-model, an energy sub-model, a non-energy intermediate input sub-model, and an oil product sub-model. The aggregate sub-model includes one output price equation and five equations relating to the cost shares of capital, labor, energy, non-energy intermediate inputs and the rate of technological change. The energy sub-model has one price (energy price) equation and four share equations explaining the cost shares of coal, oil products, natural gas, and electricity. The non-energy intermediate sub-model is composed of one price (material price) equation and five equations for the cost shares of agricultural intermediate inputs, industrial intermediate inputs, transportation's intermediate inputs, service intermediate inputs, and imported intermediate inputs. Similarly, the oil product sub-model has one price (oil price) equation and four share equations explaining the cost shares of gasoline, diesel, fuel oil and other oil products. Figure 6.1 presents the tier structure of the sub-models in the producer's model. With the sole exception of the oil sub-model, the explanatory variables consist of input prices and time as an index for the level of technology. For the oil sub-model, the explanatory variable consists of input prices only.

Taking the aggregate input sub-model as an example, the output price (P) equation is:

$$\begin{aligned}\ln P = \ln\alpha_0 + \alpha_T T + \sum_i \alpha_i \ln P_i + \frac{1}{2}\sum_i\sum_j \beta_{ij}\ln P_i \ln P_j \\ + \sum_i \beta_{iT}\ln P_i T + \frac{1}{2}\beta_{TT}T^2, \qquad i, j = K, L, E, M\end{aligned} \tag{6.1}$$

where $i, j = K, L, E, M$ denotes capital, labor, energy and intermediate inputs, respectively. T denotes time as an index for the level of technology.

The input cost share equation is:[2]

$$S_i = \alpha_i + \sum_i \beta_{ij}\ln P_i + \beta_{iT}T, \qquad i, j = K, L, E, M \tag{6.2}$$

and the rate of technical change ($-R_T$) is:

$$-R_T = \frac{\partial \ln P}{\partial T} = \alpha_T + \sum_i \beta_i \ln P_i + \beta_{TT}T \tag{6.3}$$

The basic approach of the model, which is a modification of the Hudson–Jorgenson (1974) model, is an integration of econometric modeling and input–output analysis. However, to reflect the dramatic changes in both the industrial structure and energy consumption patterns of the Taiwan economy, a time trend is included in the energy and material sub-models.

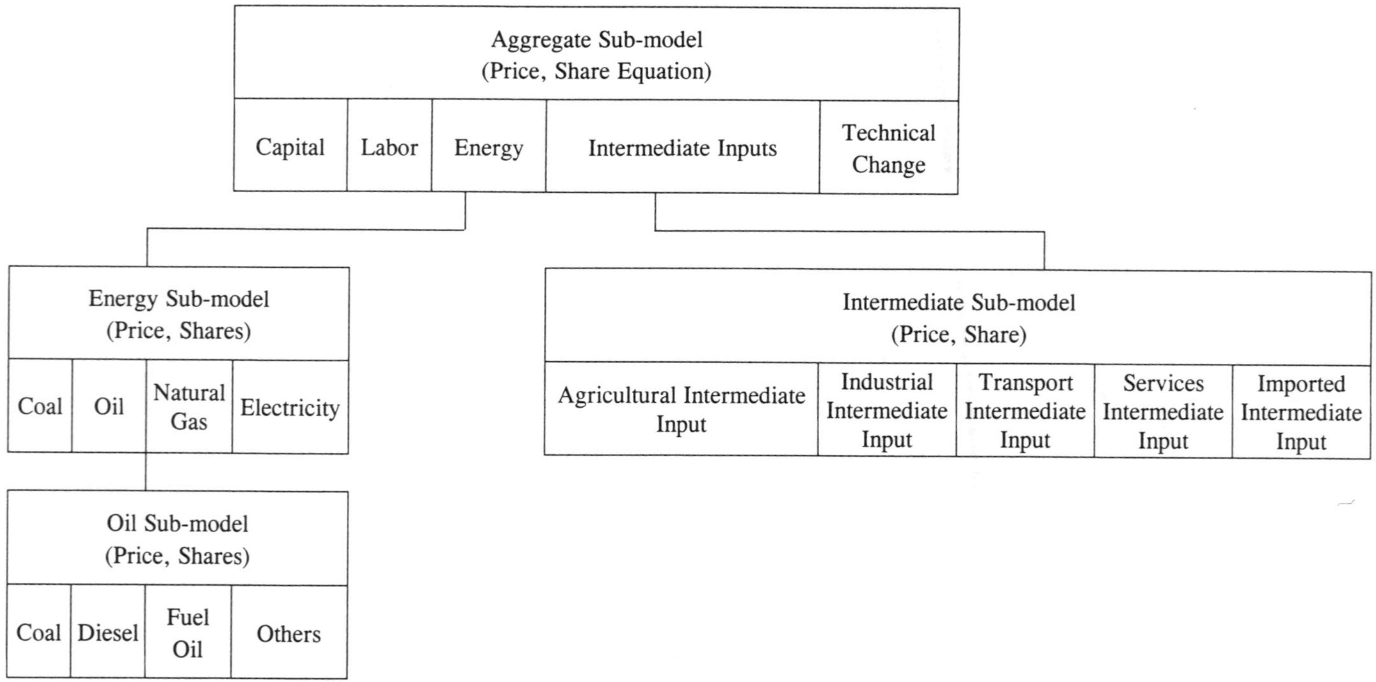

Figure 6.1 The tier structure of the producer's sub-model in the DGEMT model

This innovation makes this Jorgenson–Liang (1985) model significantly different from most of the studies by Jorgenson and his associates, which are based on highly-developed economies, such as the United States, Japan and West Germany. This kind of model will also be useful for case studies involving the other newly industrializing countries (NICs).

Liang (1987), Jorgenson and Liang (1985) and Liang (1999) contain detailed descriptions of this theoretical model, together with the estimation method, data compilation and the results of coefficients estimated. It is noted that Liang (1999) is a revised model of Jorgenson–Liang (1985) in that it updates the time-series data of the producer's model from 1961–81 to 1961–93, and also combines the consumer's model (Liang, 1983), the macroeconomic model of the Directorate-General of Budget, Accounting and Statistics, Executive Yuan (Ho et al., 2001), and the MARKAL Engineering Model of the Industrial Technology Research Institute (ITRI) (Young, 1996).

2.2 The Consumer's Model

Following Jorgenson–Slesnick (1983), we assume that the kth household allocates its expenditures in accordance with the translog indirect utility function. Under exact aggregation conditions, the vector of aggregate expenditure shares can be expressed in the following form:

$$S = \frac{1}{D(P)}\left(\alpha_p + \beta_{PP}\ln P - \beta_{PP}\cdot\iota\cdot\frac{\sum M_k \ln M_k}{M} + \beta_{PA}\frac{\sum M_k A_k}{M}\right) \tag{6.4}$$

Under exact aggregation, systems of individual expenditure shares for consuming units with identical demographic characteristics can be recovered in one and only one way from the system of aggregate expenditure shares.[3]

Equation (6.4) implies that the vector of the expenditure shares of the household sector (private consumption) are determined by commodity prices (*P*), the expenditure structure ($\sum[M_k \ln M_k / M]$) and the joint distribution of household expenditure, and the attributes ($\sum M_k A_k / M$), where M_k and A_k denote the kth household's expenditure and attributes, respectively. ι is a vector of ones. We divide private consumption into five categories:

1. Food: Expenditures on food, beverages and tobacco.
2. Clothing: Expenditures on clothing, apparel.
3. Housing: Expenditures on rent and non-energy utilities, furniture, furnishing and household equipment, household operations and services.
4. Energy: Expenditures on fuel and electricity including fuel for vehicles.
5. Recreation, Transportation and Miscellaneous: Expenditures on recreation, amusement and education, medical and health care, transportation

and miscellaneous consumption expenditures.

Hence the vector of expenditure share (S) in fact consists of the five types of expenditure share referred to above. The following demographic characteristics are employed as attributes of households:

1. Family size: 1, 2, 3, 4, 5, 6, 7, 8 or more.
2. Occupation: Non-farmer and farmer.
3. Number of persons employed: 1, 2, 3 or more.

For a detailed description of the model, please refer to Liang (1983).[4] The consumer's model is linked to the producer's model through output prices by sector, while it is linked to the DGBAS's macroeconomic model via total private consumption (see the next section).

2.3 The DGBAS Macroeconomic Model[5]

The macroeconomic model of the Directorate-General of Budget, Accounting and Statistics (DGBAS) is a Keynesian model which consists of 159 equations. We retrieve the following projection data from the macroeconomic model as initial values in the baseline projection: (1) GDP growth rate, (2) wage, (3) interest rate, (4) private consumption, (5) consumer price index (CPI), (6) wholesale price index (WPI), (7) investment, (8) government expenditure, and (9) exports. Both the CPI and WPI are affected by output prices in each sector. The GDP, wage, interest rate and private consumption are functions of the CPI or the WPI in this macroeconomic model. Thus, there are feedback relationships between the DGBAS macroeconomic model and the producer's model if sectoral output prices change due to the implementation of an energy tax.

The total supply is composed of the intermediate demands of industries and the final demands of private consumption (C), investment (I), government expenditures (G), and net exports (X) minus imports (M). Markets are cleared by the prices of domestically produced commodities for each sector (P_i).[6]

$$P_i Q_i = \sum_j P_i A_{ij} + P_i(C_i + I_i + G_i) + P_i X - P_i M_i\ ,\ i, j = 1,\ldots,29 \qquad (6.5)$$

2.4 The ITRI MARKAL Engineering Energy Model[7]

By employing the linear programming method, the ITRI MARKAL engineering model combines the information relating to the growth of industries, energy supply and energy technologies to achieve the best energy mix. This model is developed by the Institute for Energy and Resources of the Industrial Technology Research Institute (ITRI).

Because information regarding future energy technology development is given careful consideration in the model, we use the aggregate of the energy demand by types projected by the ITRI MARKAL engineering model to control for the total energy demand projected by the producer's and consumer's models.

3 SIMULATION METHODOLOGY AND PROCEDURE

The simulation framework of the model is presented in Figure 6.2.

3.1 Base Case Projection

To assess the effect of an energy tax, we must first determine the future path of the Taiwanese economy in the absence of energy tax. We call such a scenario a base case. The base case projection is conducted by means of the following steps:

1. We insert the values of the capital services price (P_K), the wage (P_L) and the price of imported intermediate inputs (P_m) projected by the DGBAS macroeconomic model into the producer's model. In this way, we obtain the prices and factor cost shares for 29 sectors over 1999–2020.
2. By employing the 1996 input–output table, we then convert the 29 sectoral output prices into the prices of five consumer goods during 1999–2020. By inserting the prices of five consumer goods together with the private consumption as projected by the macroeconomic model into the consumer's model, we obtain the shares of five consumer goods in total private consumption.
3. The demand for types of energy by sector, taking oil as an example, is derived by multiplying the oil coefficient (O/Q) by the total output (Q) for each sector. The oil coefficient (O/Q) can be calculated by means of the following equation:

$$\frac{O}{Q} = \frac{P_E \cdot E}{P \cdot Q} \cdot \frac{P_O \cdot O}{P_E \cdot E} \cdot \frac{P}{P_O} = S_E \cdot S_O \cdot \frac{P}{P_O} \tag{6.6}$$

where S_E : Energy share of total cost,

S_O : Oil share of energy cost,

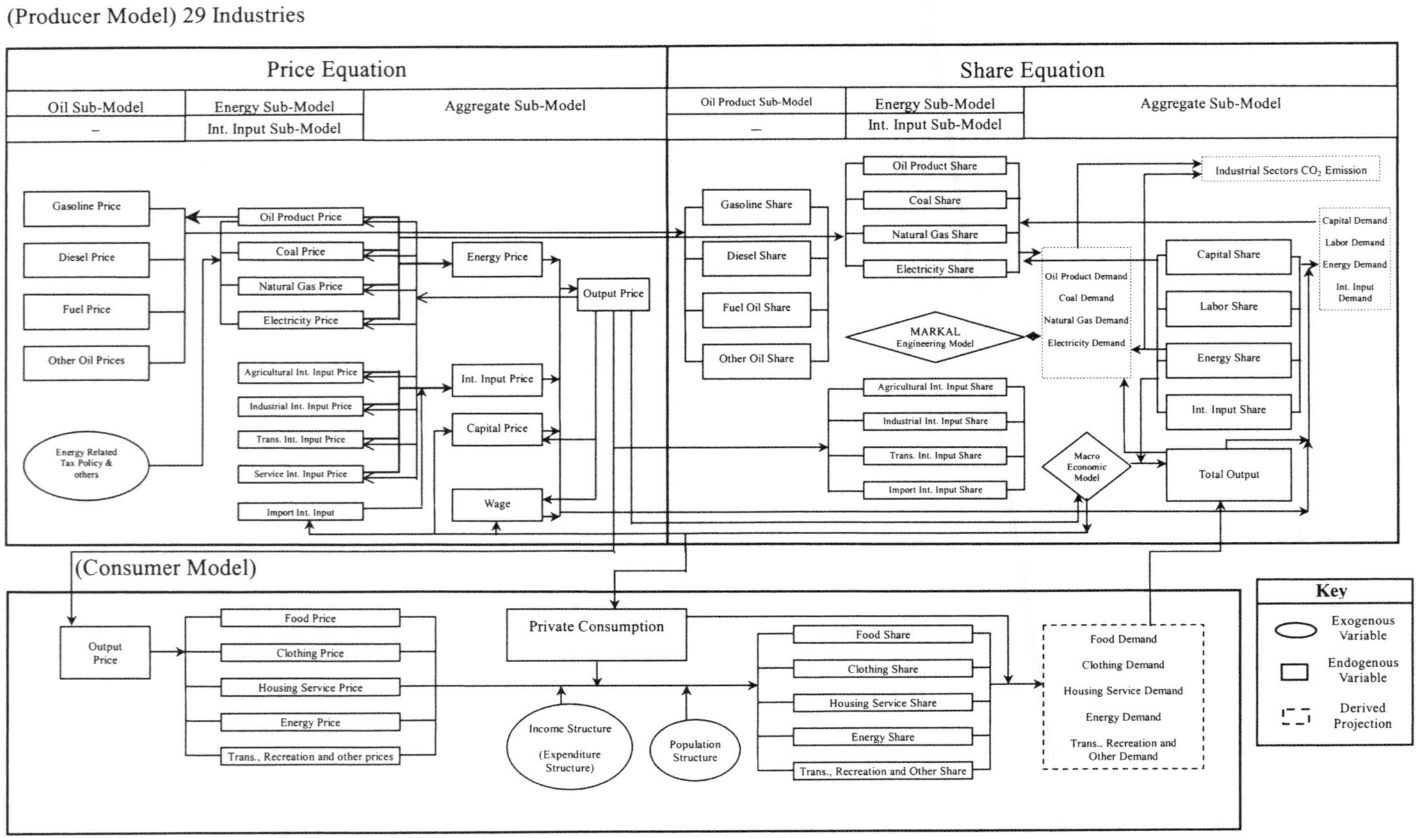

Figure 6.2 The simulation framework of the DGEMT model

P : Output price,
P_O : Price of oil products,

and S_E, S_O, P and P_O are endogenously determined in the model.

The projected growth rate of sectoral output during 1999–2020 is derived by: (i) the GDP growth rate obtained from the macroeconomic model, (ii) the industrial structure projection provided by this study, and (iii) the use of the sectoral value-added shares in total output which are endogenously determined from this model's simulation.

4. The demand for energy in the household sector (E_H) is derived by

$$E_H = S_E \cdot \frac{PC}{P_E} \tag{6.7}$$

Here, S_E, P_E and PC denote, respectively, the energy expenditure share of private consumption, the energy price and private consumption. Both S_E and P_E are determined endogenously from the consumer's model, while PC (private consumption) comes from the projection of the DGBAS macroeconomic model.

5. The demand for the various types of energy are then converted into CO_2 emissions by employing the conversion factor projected by the MARKAL engineering model, such as: coal (3.53 tons CO_2/KLOE),[8] oil products (2.89 tons CO_2/KLOE), and natural gas (2.09 tons CO_2/KLOE). This completes the whole process of baseline projection.

3.2 Simulation Involving Energy Taxes

6. Next, we evaluate the impact of an energy tax. We convert the prices of different energy types ranging from endogenous to exogenous. The prices of energy are modified by incorporating energy tax schedules into the producer's model and consumer's model, respectively, to calculate their corresponding output prices, cost shares, demand for types of energy and CO_2 emissions by sectors, as well as the consumption structure and quantity of consumer goods.
7. However, the above scenarios do not consider the 'feedback' effect in the changes in the capital service price (PK), wage (PL) and output caused by implementing the tax. In fact, the implementation of energy taxes will affect PK and PL and total output by sector as well. In the DGBAS macroeconomic model, PK and PL are affected by the energy tax through the increase in the general price level. Hence we insert the GDP deflator into the PK and PL function to obtain a new PK and PL, and in turn new values of the output price, cost structure and CO_2 emissions by sector.
8. The impact of the energy tax on total output by sector is evaluated by

means of the following procedure:

(i) First of all, we calculate the impact of the energy tax on the sectoral output price and the general price level (GDP deflator), and, in turn, the new values of final demand such as private consumption, investment, government expenditure, net exports and GDP.
(ii) Next, we multiply the private consumption by the private consumption shares of the five consumer goods, which are then deflated by their respective prices to obtain the new values of the five consumer goods.
(iii) We then employ the 1996 Input–Output table to convert the changes in the five consumer goods to the changes in sectoral final demand (*FD*).[9]
(iv) We obtain the sectoral total output (*Q*) by using the following standard input–output equation: $Q = FD \cdot (1 - D)^{-1}$. Here, *D* denotes the matrix of domestic product input–output coefficients.
(v) We calculate the energy conservation effect on the total output of the four energy sectors and the whole economy. The energy conservation effect is obtained by comparing the demand for the four types of energy in the base case with that in the 'energy tax' cases where energy taxes are implemented.

9. Finally, the impact of different energy taxes on the sectoral output price, the demand for various types of energy and CO_2 emissions are compared.

It is noted that the imposition of energy-related taxes will reduce total output and further reduce the demand for energy and CO_2 emissions. Therefore, the total impact of energy taxes on CO_2 emissions reduction should also accommodate the effect on output growth. In a nutshell, we consider not only the 'substitution effect' but also the 'income effect,' both in the consumer's and producer's models, in relation to the demand for energy and CO_2 emissions.

4 SIMULATION RESULTS

4.1 Effect of Energy Tax on Energy Prices

Based on the carbon tax schedule for Holland (US\$2.24/tons CO_2), Finland (US\$3.93/tons CO_2), Denmark (US\$14.88/tons CO_2) and Sweden (US\$ 22.2/tons CO_2), as well as a conversion factor of 0.21 tons of CO_2 per kilolitre oil equivalent (KLOE),[10] the energy tax rates are converted into the following four scenarios: NT$0.16/LOE, NT$0.28/LOE, NT$1.05/LOE, NT$1.56/LOE. LOE stands for liter oil equivalent. Among these, fuel oil has the highest tax rate, followed by coal, LPG, natural gas, diesel oil, gasoline and electricity.

Because there is not perfect substitutability among the different types of energy (for example, coal and fuel oil cannot replace gasoline and diesel for car use) and the tax rate of each kind of energy is different, the unit caloric prices of various types of energy are different in Taiwan. The present unit caloric energy price structure is as follows (take the unit caloric price of coal as 1) coal: premium gasoline: premium diesel: fuel oil: LPG: natural gas: electricity = 1: 4.25: 2.68: 0.84: 1.58: 1.63: 4.29.

By imposing the energy tax rate and taking the NT$1.56/LOE tax rate as an example (coal is NT$1.56/LOE – for other types of energy please refer to Table 6.1), the energy price structure (in NT$/LOE) will be changed to coal: premium gasoline: premium diesel: fuel oil: LPG: natural gas: electricity = 1: 3.41: 2.25: 0.88: 1.43: 1.47: 3.45 (see Table 6.1). After imposing the energy tax, each energy price related to coal declines significantly, with the exception of fuel oil. This brings advantages to natural gas, electricity and LPG that may serve as substitutes for coal and fuel oil in the producer's sector (see Table 6.1).

4.2 Effect of Energy Tax – A One-Step Approach

Effect on energy demand and CO_2 emissions

By imposing the highest energy tax (NT$1.56/LOE) through the use of a one-step approach, CO_2 emissions decrease by 20 per cent for the economy as a whole in 1999. The energy demand in relation to coal has the largest decrease, which is −25.61 per cent, followed by oil products (−18.54 per cent), natural gas (−16.31 per cent) and electricity (−12.88 per cent) (see Table 6.2).

Effect on output prices

Imposing an energy tax will of course have the greatest impact on the prices within the four energy sectors (see Table 6.3). Among manufacturing industries, the non-metallic mineral products, basic metal and chemical products industries will suffer the greatest impact in terms of price increases. Public utilities and transportation will have the highest price increases among the seven one-digit sectors. For the economy as a whole, the GDP deflator will increase by 1.79 per cent in 1999, if the highest energy tax rate of NT$1.56/LOE is imposed (see Table 6.3).

Table 6.1 Comparison of different energy taxes and energy prices in 1998

Unit: NT$

	Coal	Gasoline	Diesel	Fuel	LPG	Natural Gas	Electricity
Price in 1998	4.51 (100.00)	19.15 (100.00)	12.07 (100.00)	3.77 (100.00)	7.14 (100.00)	7.34 (100.00)	19.37 (100.00)
Converted from Dutch Carbon Tax Amount* (NT$0.16/LOE)	0.157 (3.488)	0.157 (0.821)	0.157 (1.303)	0.157 (4.172)	0.157 (2.203)	0.157 (2.143)	0.157 (0.812)
Converted from Finnish Carbon Tax Amount* (NT$0.28/LOE)	0.276 (6.119)	0.276 (1.441)	0.276 (2.286)	0.276 (7.320)	0.276 (3.865)	0.276 (3.760)	0.276 (1.425)
Converted from Danish Carbon Tax Amount* (NT$1.05/LOE)	1.045 (23.168)	1.045 (5.456)	1.045 (8.657)	1.045 (27.716)	1.045 (14.634)	1.045 (14.235)	1.045 (5.394)
Converted from Swedish Carbon Tax Amount* (NT$1.56/LOE)	1.56 (34.565)	1.56 (8.140)	1.56 (12.915)	1.56 (41.350)	1.56 (21.833)	1.56 (21.238)	1.56 (8.048)

Note: * The conversion factor from the carbon tax rate to the energy tax rate is based on Taiwan's energy usage and CO_2 emissions in 1998.

Effect on output growth

The four energy sectors will also suffer the greatest decline in output growth when the energy tax is imposed. This is due to the 'substitution effect' and 'income effect' both in terms of the final demand and the producer's sector. Similarly, the non-metallic mineral products, basic metal and chemical products sectors are among the most affected parts of the manufacturing sector. The public utility and transportation sectors will exhibit the greatest decrease in output growth among the seven one-digit sectors. For the economy as a whole, GDP will decline by 1.06 per cent if the highest energy tax rate (NT$1.56/LOE) is imposed (see Table 6.4).

Table 6.2 Effect of energy taxes on energy demand and CO_2 emissions in 1999

(NT$1.56/LOE: One-Step Approach)

Unit: %

Sector	(1) Coal	(2) Oil	(3) Natural Gas	(4) Electricity	(5) CO_2
Agriculture	–	−19.82	–	−12.88	−19.07
Mining	−20.88	−15.60	−11.61	−2.37	−19.92
Coal Mining	0.00	8.79	–	18.14	−11.09
Natural Gas	–	−3.64	0.14	5.54	−15.13
Manufacturing	−25.93	−18.44	−16.47	−13.32	−21.19
Food	−26.37	−19.88	−17.96	−12.97	−16.45
Beverages & Tobacco	−26.61	−20.14	−18.23	−13.25	−18.58
Textiles	−25.48	−18.01	−16.98	−12.13	−14.68
Clothes & Wearing Apparel	−26.53	−19.50	–	−13.39	−17.08
Leather & Leather Products	−27.28	−11.17	485.97	−16.78	−15.52
Wood & Bamboo Products	–	−19.47	−17.54	−12.53	−15.56
Furniture Products	–	–	–	–	–
Paper & Printing	−25.48	−18.92	−16.97	−11.92	−15.56
Chemicals & Plastics	−60.56	−16.76	−85.30	−7.63	−19.55
Rubber Products	−25.99	−19.62	–	−12.47	−16.48
Oil Refining	−8.08	−0.16	9.31	3.27	−18.91
Non-Metallic Minerals	−24.40	−17.96	−7.14	−12.96	−22.84
Basic Metals	−34.11	−13.51	−15.56	−6.69	−27.22
Metal Products	−25.85	−19.31	−17.38	−12.36	−16.80
Machinery & Equipment	−25.69	−19.14	−7.74	−12.29	−17.17
Elect. Mach. & Electronics	−26.73	−20.27	−18.37	−13.40	−15.96
Transportation Equipment	−26.99	−20.55	−18.65	−13.70	−17.44
Miscellaneous	−100.86	−100.86	−100.86	−100.86	−107.63
Water, Electricity & Gas	−17.85	−18.99	−12.02	−5.09	−25.54
Electricity	−10.33	−16.61	−4.45	2.53	−22.53
Construction	−24.94	−18.33	–	−11.29	−22.76
Transportation & Comm.	−25.01	−18.40	–	−11.42	−19.85
Services	−25.76	−19.22	−17.29	−12.26	−13.96
Industry	−25.61	−18.46	−16.30	−13.14	−21.61
Whole Economy	−25.61	−18.54	−16.31	−12.88	−20.00

Table 6.3 Effect of different energy taxes on prices by sector in 1999

(One-Step Approach)

Unit: %

Sector	Holland Tax (NT$0.16/ LOE)	Finland Tax (NT$0.28/ LOE)	Denmark Tax (NT$1.05/ LOE)	Sweden Tax (NT$1.56/ LOE)
Agriculture	0.14	0.24	0.92	1.37
Mining	0.98	1.73	6.54	9.75
Coal Mining	3.75	6.59	24.94	37.21
Natural Gas	2.35	4.12	15.61	23.30
Manufacturing	0.20	0.34	1.30	1.94
Food	0.14	0.24	0.90	1.34
Beverages & Tobacco	0.10	0.18	0.68	1.02
Textiles	0.26	0.45	1.71	2.54
Clothes & Wearing Apparel	0.17	0.31	1.16	1.73
Leather & Leather Products	0.09	0.16	0.59	0.88
Wood & Bamboo Products	0.14	0.24	0.90	1.35
Furniture Products	–	–	–	–
Paper & Printing	0.26	0.45	1.71	2.55
Chemicals & Plastics	0.29	0.51	1.95	2.91
Rubber Products	0.19	0.33	1.24	1.85
Oil Refining	2.64	4.64	17.57	26.21
Non-Metallic Minerals	0.41	0.71	2.69	4.02
Basic Metals	0.33	0.58	2.19	3.26
Metal Products	0.21	0.36	1.37	2.05
Machinery & Equipment	0.17	0.30	1.14	1.70
Elect. Mach. & Electronics	0.10	0.17	0.64	0.96
Transportation Equipment	0.12	0.22	0.82	1.23
Miscellaneous	0.15	0.26	0.99	1.47
Water, Electricity & Gas	1.38	2.42	9.17	13.68
Electricity	1.64	2.89	10.92	16.30
Construction	0.22	0.38	1.45	2.16
Transportation & Comm.	0.33	0.58	2.20	3.28
Services	0.10	0.18	0.67	1.01
Industry	0.29	0.51	1.93	2.88
GDP Deflator	0.18	0.32	1.20	1.79

Table 6.4 Effect of energy taxes on output growth in 1999

(One-Step Approach)

Unit: %

Sector	Holland Tax (NT$0.16/ LOE)	Finland Tax (NT$0.28/ LOE)	Denmark Tax (NT$1.05/ LOE)	Sweden Tax (NT$1.56/ LOE)
Agriculture	−0.14	−0.24	−0.92	−1.22
Mining	−0.45	−0.78	−2.96	−3.92
Coal Mining	−3.38	−5.77	−18.67	−22.66
Natural Gas	−1.96	−3.39	−11.54	−14.43
Manufacturing	−0.16	−0.29	−1.10	−1.46
Food	−0.04	−0.07	−0.26	−0.33
Beverages & Tobacco	−0.08	−0.13	−0.45	−0.57
Textiles	−0.09	−0.15	−0.53	−0.67
Clothes & Wearing Apparel	−0.07	−0.12	−0.42	−0.54
Leather & Leather Products	−0.06	−0.10	−0.34	−0.43
Wood & Bamboo Products	−0.09	−0.15	−0.52	−0.66
Furniture Products	−0.09	−0.15	−0.52	−0.65
Paper & Printing	−0.12	−0.21	−0.74	−0.94
Chemicals & Plastics	−0.21	−0.37	−1.27	−1.59
Rubber Products	−0.08	−0.14	−0.49	−0.62
Oil Refining	−2.28	−3.92	−13.22	−16.40
Non-Metallic Minerals	−0.34	−0.58	−2.01	−2.54
Basic Metals	−0.70	−1.15	−2.82	−2.71
Metal Products	−0.07	−0.12	−0.41	−0.52
Machinery & Equipment	−0.06	−0.11	−0.37	−0.47
Elect. Mach. & Electronics	−0.04	−0.07	−0.22	−0.28
Transportation Equipment	−0.04	−0.06	−0.22	−0.27
Miscellaneous	−0.27	−1.32	−4.54	−5.75
Water, Electricity & Gas	−1.21	−2.09	−7.78	−10.12
Electricity	−1.47	−2.54	−8.95	−11.40
Construction	−0.09	−0.16	−0.61	−0.81
Transportation & Comm.	−0.22	−0.39	−1.48	−1.97
Services	−0.05	−0.08	−0.30	−0.40
Industry	−0.23	−0.40	−1.52	−2.00
GDP	−0.12	−0.21	−0.80	−1.06

Table 6.5 Time schedule for 22-year progressive ad valorem *energy tax rate*

(NT$1.56/LOE)

Unit: %

Year	Coal	Gasoline	Diesel	Fuel	LPG	Natural Gas	Electricity
1999	1.3586	0.3563	0.5536	1.5855	0.9017	0.8792	0.3525
2000	2.7356	0.7140	1.1103	3.1961	1.8115	1.7662	0.7062
2001	4.1314	1.0728	1.6701	4.8322	2.7295	2.6610	1.0611
2002	5.5461	1.4330	2.2330	6.4943	3.6558	3.5636	1.4173
2003	6.9800	1.7945	2.7990	8.1828	4.5904	4.4741	1.7748
2004	8.4335	2.1572	3.3682	9.8980	5.5335	5.3927	2.1335
2005	9.9066	2.5212	3.9405	11.6404	6.4851	6.3194	2.4935
2006	11.3998	2.8866	4.5159	13.4104	7.4453	7.2541	2.8547
2007	12.9133	3.2532	5.0946	15.2085	8.4141	8.1972	3.2173
2008	14.4473	3.6211	5.6764	17.0351	9.3916	9.1485	3.5811
2009	16.0022	3.9904	6.2615	18.8907	10.3780	10.1081	3.9461
2010	17.5781	4.3609	6.8498	20.7757	11.3732	11.0762	4.3125
2011	19.1755	4.7328	7.4413	22.6906	12.3775	12.0528	4.6802
2012	20.7947	5.1060	8.0362	24.6358	13.3908	13.0381	5.0491
2013	22.4358	5.4806	8.6343	26.6119	14.4132	14.0319	5.4194
2014	24.0992	5.8565	9.2358	28.6193	15.4448	15.0345	5.7910
2015	25.7851	6.2337	9.8405	30.6585	16.4858	16.0459	6.1638
2016	27.4940	6.6122	10.4487	32.7301	17.5361	17.0663	6.5380
2017	29.2262	6.9921	11.0601	34.8345	18.5959	18.0955	6.9135
2018	30.9818	7.3734	11.6750	36.9722	19.6653	19.1339	7.2903
2019	32.7613	7.7560	12.2933	39.1439	20.7443	20.1813	7.6685
2020	34.5650	8.1400	12.9150	41.3500	21.8330	21.2380	8.0480

4.3 Effect of Energy Tax – A Progressive Approach

According to the above simulation, we found the impact of energy tax on the economy to be significant if the tax rate is as high as the Swedish tax rate (NT$1.56/LOE). Alternatively, we might choose a progressive approach to achieve the same goal of CO_2 reduction, while minimizing its impact on the economy. Using a progressive *ad valorem* tax approach, we assume that the tax rate in 2020 is the same as that when a one-step approach is used in 1999. The schedule for the 22-year progressive *ad valorem* tax rate (converted from the Swedish carbon tax amount) is presented in Table 6.5.

Effect on energy demand and CO_2 emissions

By imposing the highest energy tax (NT$1.56/LOE) on the 22-year progressive tax rate, CO_2 emissions will decline by 20.00 per cent by 2020 for the economy as a whole (see Table 6.6). This is almost the same as the rate of reduction when employing a one-step approach for 1999.

The energy demanded in relation to coal has the largest decrease, which is −23.13 per cent, followed by oil products (−19.95 per cent), natural gas (−17.55 per cent), and electricity (−12.27 per cent).

Effect on prices

Using the progressive approach to implement the highest energy tax rate (NT$1.56/LOE), the effect of the energy tax on prices in each sector is shown in Table 6.7.

By comparing Table 6.3 with Table 6.7, we found that the progressive tax approach can effectively reduce the negative effect on the price level. For instance, at the same energy tax rate of NT$1.56/LOE, using the one-step approach will increase the GDP deflator by 1.79 per cent, while using the 22-year progressive approach will increase the GDP deflator by 0.81% per cent, or less than half the increase from using the one-step approach.

By imposing the 22-year progressive approach, the water, electricity and gas sector (12.18 per cent) is affected the most in terms of a price increase among the seven one-digit sectors. It is followed by mining (9.05 per cent), construction (1.38 per cent), manufacturing (1.37 per cent), transportation (0.92 per cent) and agriculture (0.64 per cent). The top five manufacturing sectors in terms of respective price increases are oil refining (26.24 per cent), non-metallic minerals (2.89 per cent), basic metals (2.62 per cent), chemicals and plastics (2.04 per cent) and textiles (1.58 per cent).

To sum up, sectoral ranking in terms of the impact of price increases as a result of imposing the 22-year progressive energy tax is identical with the case where the one-step energy tax is used (see Tables 6.7 and 6.3).

Table 6.6 Effect of energy taxes on energy demand and CO_2 emissions by 2020

(NT$1.56/LOE: Progressive Approach)

Unit: %

Sector	(1) Coal	(2) Oil	(3) Natural Gas	(4) Electricity	(5) CO_2
Agriculture	-	−20.52	-	−11.95	−18.39
Mining	−18.76	−16.63	−12.38	−0.43	−19.51
Coal Mining	0.00	4.85	-	15.83	−10.89
Natural Gas	−100.81	11.21	−0.44	19.21	−8.86
Manufacturing	−23.57	−19.04	−16.99	−12.22	−20.67
Food	−24.16	−20.48	−18.50	−12.13	−14.70
Beverages & Tobacco	−24.27	−20.59	−18.64	−12.25	−17.58
Textiles	−23.41	−18.14	−17.10	−11.48	−14.07
Clothes & Wearing Apparel	−23.97	−19.50	-	−12.15	−16.17
Leather & Leather Products	−24.47	−20.80	904.67	−12.49	−14.12
Wood & Bamboo Products	−15.96	−20.06	−18.08	−11.67	−15.85
Furniture Products	-	-	-	-	-
Paper & Printing	−23.32	−19.61	−17.59	−11.16	−14.40
Chemicals & Plastics	−100.81	−19.53	−99.42	−10.37	−18.10
Rubber Products	−9.26	-	-	−11.66	−12.29
Oil Refining	−5.21	0.07	−4.39	8.43	−20.31
Non-Metallic Minerals	−22.42	−18.76	−7.46	−11.88	−21.65
Basic Metals	−20.58	−23.00	−23.40	−20.37	−40.02
Metal Products	−23.57	−19.86	−17.87	−11.45	−16.09
Machinery & Equipment	−23.82	−20.11	−8.13	−11.83	−17.56
Elect. Mach. & Electronics	−24.40	−20.73	−18.78	−12.40	−15.12
Transportation Equipment	−39.41	−36.46	−33.09	−29.77	−33.12
Miscellaneous	−23.73	−20.03	−20.03	−11.63	-
Water, Electricity & Gas	−17.73	−14.74	−9.34	−6.09	−22.90
Electricity	−12.60	−9.78	−2.62	−0.99	−22.24
Construction	−23.21	−19.48	-	−11.03	−22.15
Transportation & Comm.	−23.94	−20.23	-	−12.04	−22.31
Services	−25.00	−21.36	−19.21	−13.10	−14.99
Industry	−23.13	−18.92	−16.71	−11.95	−20.92
Whole Economy	−23.13	−19.95	−17.55	−12.27	−20.00

Table 6.7 Effect of different energy taxes on prices by sector by 2020

(Progressive Approach)

Unit: %

Sector	Denmark Tax (NT$1.05/LOE)	Sweden Tax (NT$1.56/LOE)
Agriculture	0.44	0.64
Mining	6.09	9.05
Coal Mining	21.67	32.31
Natural Gas	24.92	37.18
Manufacturing	0.92	1.37
Food	0.41	0.60
Beverages & Tobacco	0.31	0.46
Textiles	1.07	1.58
Clothes & Wearing Apparel	0.75	1.11
Leather & Leather Products	0.45	0.66
Wood & Bamboo Products	0.67	0.99
Furniture Products	–	–
Paper & Printing	1.15	1.70
Chemicals & Plastics	1.38	2.04
Rubber Products	0.81	1.20
Oil Refining	17.59	26.24
Non-Metallic Minerals	1.93	2.89
Basic Metals	1.76	2.62
Metal Products	0.93	1.37
Machinery & Equipment	0.73	1.09
Elect. Mach. & Electronics	0.40	0.58
Transportation Equipment	0.52	0.76
Miscellaneous	0.78	1.16
Water, Electricity & Gas	8.17	12.18
Electricity	9.65	14.36
Construction	0.93	1.38
Transportation & Comm.	0.67	0.92
Services	–	–
Industry	1.49	2.22
GDP Deflator	0.55	0.81

Table 6.8 Effect of energy taxes on output growth by 2020

(Progressive Approach)

Unit: %

Sector	Denmark Tax (NT$1.05/LOE)	Sweden Tax (NT$1.56/LOE)
Agriculture	−1.03	−1.16
Mining	−2.72	−3.02
Coal Mining	−16.73	−17.22
Natural Gas	−12.40	−13.06
Manufacturing	−0.93	−1.05
Food	−0.25	−0.26
Beverages & Tobacco	−0.46	−0.49
Textiles	−0.67	−0.71
Clothes & Wearing Apparel	−0.50	−0.53
Leather & Leather Products	−0.44	−0.47
Wood & Bamboo Products	−0.78	−0.83
Furniture Products	−0.76	−0.81
Paper & Printing	−0.59	−0.62
Chemicals & Plastics	−0.97	−1.03
Rubber Products	−0.30	−0.32
Oil Refining	−14.20	−14.85
Non-Metallic Minerals	−1.57	−1.67
Basic Metals	−3.00	−2.42
Metal Products	−0.35	−0.38
Machinery & Equipment	−0.46	−0.49
Elect. Mach. & Electronics	−0.32	−0.34
Transportation Equipment	−0.05	−0.05
Miscellaneous	–	–
Water, Electricity & Gas	−6.12	−6.76
Electricity	−8.50	−9.13
Construction	−0.79	−0.88
Transportation & Comm.	−1.79	−2.01
Services	−0.27	−0.31
Industry	−1.29	−1.44
GDP	−0.73	−0.82

Effect on Output Growth
By comparing Table 6.8 with Table 6.4, we conclude that, when the 22-year progressive energy tax is imposed, GDP will be reduced by 0.82 per cent by 2020, which is less than the reduction from using a one-step approach (−1.19 per cent). Similarly, the sectoral ranking in terms of the decrease in output as a result of imposing the 22-year progressive energy tax is also the same as that in the case of the one-step energy tax.

4.4 Marginal Social Abatement Cost of CO_2 Emissions

Here, we define the marginal social abatement cost of CO_2 emissions as the change in GDP divided by the change in CO_2 emissions. The figures denoting the changes in both GDP and CO_2 emissions are derived from various scenarios related to the energy tax schemes mentioned above.

Table 6.9 presents the marginal social abatement cost of CO_2 reductions by employing various tax rates in 1999. We find that the marginal social abatement cost rises significantly with the increase in the targeted CO_2 reduction. For instance, when the targeted CO_2 reduction increases from 2.48 per cent to 20 per cent, the marginal social abatement cost will increase from NT$2107/ton to NT$2572/ton in 1999, an 18.08 per cent increase.

To compare the abatement cost of CO_2 emission by using 'one-step' and 'progressive' approach, a time path analysis is required. First, we estimate the impact of energy tax on CO_2 emission and GDP year by year during 1999–2020. Second, a 5.38 per cent discounted rate, which is the average government bond rate (ten years) during 1995–2003, is employed to discount the change of GDP at 1999 constant prices to the present value.

Taking the highest tax rate of (NT$1.56/LOE) as an example, Tables 6.10 and 6.11 present marginal social abatement cost of CO_2 reduction during 1999–2020 by adopting 'one-step' approach and 'progressive' approach respectively.

We conclude that the average of marginal social abatement cost of CO_2 reduction during 1999–2020 is NT$1032 per ton by 'progressive' approach (See Table 6.11), which is 31.3 per cent lower than that of 'one-step' approach (NT$1502 per ton) (See Table 6.10).

5 CONCLUSION AND IMPLICATIONS

From the above findings, we conclude that relying on an energy tax to lower CO_2 emissions by as much as 20 per cent will have a significant effect on economic growth, the inflation rate as well as the marginal social abatement cost of CO_2 emissions, if a one-step approach is applied. However, implementing a 'progressive' energy tax will lessen the negative effect on the

Table 6.9 Marginal social abatement cost of CO2 reduction in Taiwan, 1999

Energy Tax Rate	(1) CO_2 Reduction Rate (%)	(2) Change in CO_2 (Million Tons)	(3) GDP Change (%)	(4) Change in GDP at 1999 Prices (Million NT$)	(5)=(4)/(2) MSAC at 1999 Prices (NT$/Ton)
NT$0.16/LOE	−2.48	−5.194	−0.12	−10946	2107
NT$0.28/LOE	−4.26	−8.933	−0.21	−19174	2146
NT$1.05/LOE	−14.34	−30.045	−0.80	−72218	2404
NT$1.56/LOE	−20.00	−42.000	−1.06	−95912	2572

Notes:

1. CO_2 emissions in 1999 (base case) are obtained from the projection of MARKAL, provided by Ren-Tseng Young.
2. The GDP deflator projected by the DGBAS macroeconomic model is employed to convert the imputed abatement cost at 2020 prices into 1999 prices.
3. Energy tax rates of NT$0.16/LOE and NT$0.28/LOE are too small for adopting a progressive *ad valorem* tax. Consequently, their simulations are omitted.
4. The marginal social abatement cost of CO_2 at 1999 present value is discounted using the interest rate during 1999-2020. The interest rate during 1999–2020 is obtained from the DGBAS macroeconomic model.

Table 6.10 Marginal social abatement cost of CO_2 reduction in Taiwan, 1999–2020, NT$1.56/LOE, one-step approach

Year	(1) CO_2 Reduction Rate	(2) Change of CO_2	(3) GDP Change	(4) Change of GDP at 1999 Prices	(5) Change of GDP (present value)	(6)=(4)/(2) MSAC	(7)=(5)/(2) MSAC (present value)
	(%)	(Million Tons)	(%)	(Million NT$)	(Million NT$)	(NT$/Ton)	(NT$/Ton)
1999	−20.00	−42.00	−1.06	−95912	−95912	2572	2572
2000	−19.04	−41.32	−1.01	−95668	−90784	2608	2475
2001	−18.08	−40.66	−0.96	−95448	−85951	2645	2381
2002	−17.33	−40.38	−0.92	−95393	−81516	2661	2274
2003	−16.62	−40.11	−0.87	−95362	−77329	2678	2172
2004	−15.94	−39.86	−0.83	−95353	−73374	2695	2074
2005	−15.29	−39.62	−0.79	−95368	−69639	2712	1980
2006	−14.68	−39.39	−0.76	−95404	−66108	2729	1891
2007	−14.09	−39.17	−0.72	−95462	−62772	2745	1805
2008	−13.53	−38.97	−0.69	−95543	−59618	2762	1724
2009	−12.99	−38.77	−0.66	−95646	−56635	2779	1646
2010	−12.48	−38.59	−0.63	−95770	−53813	2796	1571
2011	−11.99	−38.39	−0.60	−95902	−51136	2814	1501
2012	−11.58	−38.44	−0.57	−96161	−48657	2818	1426
2013	−11.06	−38.02	−0.55	−96229	−46205	2851	1369
2014	−10.63	−37.85	−0.52	−96424	−43935	2870	1308
2015	−10.22	−37.70	−0.50	−96642	−41786	2888	1249
2016	−9.82	−37.55	−0.48	−96880	−39751	2906	1192
2017	−9.45	−37.42	−0.45	−97140	−37823	2924	1139
2018	−9.09	−37.30	−0.43	−97422	−35996	2942	1087
2019	−8.75	−37.20	−0.42	−97726	−34265	2960	1038
2020	−8.42	−37.10	−0.40	−98051	−32624	2977	991
Average	−13.23	−38.90	−0.67	–	−58438	–	1502

Table 6.11 Marginal social abatement cost of CO_2 reduction in Taiwan, 1999–2020, NT\$1.56/LOE, progressive approach

Year	(1) CO_2 Reduction Rate	(2) Change of CO_2	(3) GDP Change	(4) Change of GDP at 1999 Prices	(5) Change of GDP (present value)	(6)=(4)/(2) MSAC	(7)=(5)/(2) MSAC (present value)
	(%)	(Million Tons)	(%)	(Million NT\$)	(Million NT\$)	(NT\$/Ton)	(NT\$/Ton)
1999	−1.05	−2.21	−0.05	−4656	−4656	2107	2107
2000	−1.84	−3.99	−0.09	−8317	−7892	2084	1978
2001	−2.62	−5.88	−0.12	−12307	−11083	2093	1885
2002	−3.43	−7.99	−0.16	−16701	−14271	2089	1785
2003	−4.26	−10.29	−0.20	−21498	−17433	2090	1694
2004	−5.11	−12.77	−0.23	−26731	−20569	2093	1610
2005	−5.97	−15.46	−0.27	−32434	−23684	2098	1532
2006	−6.84	−18.36	−0.31	−38635	−26772	2104	1458
2007	−7.73	−21.49	−0.34	−45379	−29839	2111	1388
2008	−8.63	−24.86	−0.38	−52693	−32880	2120	1323
2009	−9.54	−28.48	−0.42	−60624	−35897	2129	1261
2010	−10.47	−32.36	−0.45	−69210	−38889	2139	1202
2011	−11.39	−36.47	−0.49	−78483	−41848	2152	1147
2012	−12.42	−41.22	−0.53	−88652	−44857	2151	1088
2013	−13.26	−45.57	−0.56	−99309	−47684	2179	1046
2014	−14.20	−50.58	−0.60	−110969	−50562	2194	1000
2015	−15.15	−55.91	−0.64	−123533	−53414	2209	955
2016	−16.11	−61.59	−0.67	−137058	−56236	2225	913
2017	−17.08	−67.63	−0.71	−151611	−59031	2242	873
2018	−18.05	−74.05	−0.75	−167253	−61797	2259	835
2019	−19.02	−80.85	−0.78	−184056	−64534	2276	798
2020	−20.00	−88.08	−0.82	−202098	−67242	2295	763
Average	−10.19	−35.73	−0.43	–	−36867	–	1032

Note: The average of government bond rate (10 years) during 1995–2003 is employed as the discount rate.

economy. Consequently, it is recommended that the 'progressive' energy tax approach instead of the 'one-step' energy tax approach be adopted.

It is recognized that the effect of tax revenue is not discussed here. In fact, this chapter implicitly assumes that the tax revenue is used to reduce the government deficit. If the tax revenue can be used to reduce distortionary taxes elsewhere in the economy, such as income tax, the impact of the energy tax on the economy will be reduced.[11] This deserves further study in the future.

ACKNOWLEDGEMENTS

Chi-Yuan Liang would like to thank Professor Mun-sing Ho of Harvard University, Professor M. Kuroda and Professor H. Sakuramoto of Keio University, Mr Chin-sheun Ho of DGBAS, Mr Ren-tseng Young of ITRI, Fang-yi Lin of CEPD and one anonymous referee, for their valuable advice and comments. He is also indebted to Ms Ting Lie, Ms Wen-Ting Chen, Mr Chih-Chun Liu and Ms Yen-ling Lin for their capable assistance in data compilation, calculation and typing. The research grant from Environmental Protection Administration is also gratefully acknowledged.

NOTES

1. For the evaluation of carbon tax on Taiwan's economy, please refer to Liang (2002).
2. Based on Shephard's lemma, the input cost share equation (S_i) can be derived by differentiating equation (6.1) with the logarithmic form of the price of input (P_i).
3. See Jorgenson and Slesnick (1983).
4. It is noted that although the Liang (1987) model is more up-to-date, in that model the monotonicity constraint is not imposed, so that it might lead to the indirect utility function not being well-behaved. Therefore, we use the Liang (1983) model here instead.
5. Please refer to Ho et al. (2001).
6. Please refer to Ho (2000).
7. Please refer to Young (1996).
8. KLOE stands for kiloliter oil equivalent.
9. Here we assume that the rest of the final demand by sector changes in the same way as private consumption.
10. The primary energy consumption was 923450 thousand kiloliters and the total amount of CO_2 emissions was 196498 thousand tons in 1998. Hence we employ the 0.21 tons of CO_2/KLOE as a conversion factor, which is calculated from 196498 thousand tons of CO_2/923450 thousand KLOE.
11. For an empirical study on the impact of carbon tax on the economy if tax revenue is used to reduce distortionary taxes in the United States, please refer to Jorgenson and Wilcoxen (1993), pp. 20–23.

REFERENCES

Ho, Chin-sheun, Jeff Chien-fu Lin and Jan-sho Wang (2001), 'The Econometric Analysis Trade Relationship in the Asia-Pacific Area', International Conference on Trade, Investment, and Industrial Policy in the Asia-Pacific Region.

Ho, Mun S. (2000), 'Modeling Trade Policies and US Growth: Some Methodological Issues', in Lawrence J. Lau (ed.), *Econometrics*, MIT Press, Chapter 12, **2**, pp. 327–76.

Hudson, E.A. and D.W. Jorgenson (1974), 'US Energy and Economic Growth, 1975–2000', *The Bell Journal of Economics*, No. 2, Autumn.

Jorgenson, D.W. and Chi-Yuan Liang (1985), 'A Study on the Energy–Economic Model of Taiwan', Project report submitted to the Energy Committee, Ministry of Economic Affairs.

Jorgenson, D.W. and D.T. Slesnick (1983), 'Individual and Social Cost-of-Living Indexes', in D.W. Diewert and C. Montmarquette (eds), *Price Level Measurement*, Ottawa: Statistics Canada.

Jorgenson, D.W. and P.J. Wilcoxen (1993), 'Reducing US Carbon Emissions: An Econometric General Equilibrium', *Resource and Energy Economics*, **15** (1), pp. 7–25.

Liang, Chi-Yuan (1983), 'A Study on the Translog Model of Aggregate Consumer Demand for Energy in Taiwan', *Academia Economic Papers*, **11** (2), Institute of Economics, Academia Sinica, pp. 167–218.

Liang, Chi-Yuan (1987), 'A Study on the Energy–Economic Model of Taiwan', *Studies on the Modern Economy Series*, No. 7 (June), Institute of Economics, Academia Sinica.

Liang, Chi-Yuan (1990), 'Impact of Energy Pricing Policies on Consumption Patterns and Household Economic Welfare in Taiwan', Discussion Paper, No. 9007 (July), Institute of Economics, Academia Sinica.

Liang, Chi-Yuan (1999), 'Impact of Carbon Policy on CO_2 and Economic Development in Taiwan, 1997–2010', Paper submitted to the Environmental Protection Administration, July.

Liang, Chi-Yuan (2002), 'Effect of Carbon Tax and Energy Tax on CO_2 Emission and Economic Development of Taiwan, 1999–2020', Mimeograph.

Young, Ren-Tseng (1996), 'Adjustment in Energy Policy to Cope with the Climate Change Proposal', Project report submitted to the Energy Commission.

7. Impact of carbon tax and reduced CO_2 emissions on the Chinese economy: a static CGE analysis

Yuxin Zheng and Gang Ma

1 INTRODUCTION

Global climatic warming is a common challenge facing mankind, and also one of the global ecological crises that is most concerning mankind. Such concern has been fully reflected in the Framework Convention on Climatic Changes (FCCC) signed by more than 150 countries, and several subsequent negotiations. Meantime, in view of the ambiguity in the statements of the framework convention on key problems, and the intensive quarrels emerging in later negotiations, it could also be seen that genuine cooperation among mankind across national boundaries is very difficult.

How to reduce the emission of greenhouse gases is the central problem facing the FCCC. Carbon dioxide produced by combustion of fossil fuels is the principal contributor to the build-up of greenhouse gases, so the allocation of its reduction of emissions has become the focus of negotiations about FCCC fulfillment.

China is one of the signatories of the FCCC. According to the rules of the FCCC, China, as a developing country, is under no obligation to reduce CO_2 emissions. However, it does not mean that China can neglect her own efforts to mitigate the process of global warming. Like most developing countries, China has to face global environmental problems, while serious domestic environmental problems remain to be solved. As a responsible member of the world community, China is greatly concerned about the global environment. As a result of rapid economic growth, China has become the second largest energy consumer after the US. Fossil fuels dominate the energy structure of China. The use of fossil fuels comprises 94 per cent of China's total energy usage.

With the rapid economic growth, CO_2 emissions have also been growing rapidly in China. Between 1986 and 1996, the net increase of CO_2 emissions in China was the highest in the world. It is anticipated that between 2010 and 2020 emissions will exceed those of the US, and China will become the largest producer of CO_2 emissions in the world. Although CO_2 emissions averaged per capita in China are much lower than those of the US, China's potential impact on world climatic changes should not be neglected.

China, as a low income country, whose per capita CO_2 emissions are also low, cannot reduce CO_2 emissions by restraining economic development. In addition, economic development itself is a fundamental means to reduce CO_2 emissions. In fact, as a result of more than a decade of rapid economic development, China has been able to introduce a series of advanced energy-saving technologies and techniques promoting greater efficiency. Hence, while the Chinese economy is growing rapidly, energy consumption has grown at a relatively slower pace. Since China's economic reform in 1979, CO_2 emissions have been reduced cumulatively by 18.3 billion tons.

Many people are involved with the policy of CO_2 emission reduction in China. The key problem is the comparison of cost and benefits, that is to say, whether benefits accrued to emission reduction could compensate for the costs incurred. There is no reason for China to refuse those emission reducing measures that do not affect economic development. Recently, some scholars in developed countries used Computable General Equilibrium (CGE) models to analyse the costs of emission reduction in China, and obtained various results. The purpose of this study is to obtain more reliable and coherent results through our own research.

2 PREVIOUS RESEARCH RESULTS

Several earlier studies on CO_2 emissions in China will be reviewed in this section. First, in 1994, the World Bank (1994) conducted a study about the emission of greenhouse gases in China as of 2020. It was quite comprehensive, including every major greenhouse gas and its source, such as methane generated by rice planting. The methodology is a combination of the top-down and bottom-up methods. It used a macro-econometric model to derive final demands for 18 sectors under different growth rates; it then used these final demands along with an input–output coefficient matrix based on international experience to forecast the industrial structure of China in 2010 and 2020. Based on this information, it computed the CO_2 produced by the economic activities. With the bottom-up method, alternative energy-saving techniques in energy-intensive sectors could be studied in detail under different growth rates. On introduction of these techniques, the direct consumption coefficients in the input–output table would decrease. Hence, we have no difficulty in grasping one of its most important conclusions: a higher economic growth rate does not mean a higher level of CO_2 emissions, since, as the economy grows, it is possible to introduce more effective technology.

In other words, this study aimed to predict the future level of CO_2 emissions in China by emphasizing the role of technology selection without any economic means to control CO_2 emission, and also without any cost analysis of CO_2 emission reduction.

In a similar way to our study, Zhang (1996) and Jorgenson et al. (1998) used their respective dynamic CGE model of the Chinese economy to investigate the effects of carbon taxation. The principal reason for using CGE models to analyse carbon taxation was that the carbon tax would change the relative prices among commodities; meanwhile it would exert a substantial influence on the economy because of the intrinsic interdependence in economic structure. Both models were dynamic so that they could derive economic growth and channels of CO_2 emissions endogenously. Thus, CO_2 emission could be reduced without corresponding policies on carbon taxation.

In respect to model structure, the model of Jorgenson et al. (1998) was much more complex than Zhang's. One of its characteristics was that it considered the rent distribution arising from the dual pricing mechanism and its impact on investment. It also considered the drawbacks in investment allocation caused by an imperfect capital market in China. In contrast, Zhang's model was a typical recursive dynamic CGE model. One contribution of his work was to link the CGE model with the energy technology selection model, MARKAL. While it investigated the effects of emission reduction, the minimum-cost technology of CO_2 emission reduction was also shown. The combination of two types of modeling means that the top-down and bottom-up approaches can be synthesized to evaluate in evaluating the emission costs. Such synthesis has been far from perfect till now. Results of the CGE model could be used as inputs to the MARKAL model. However, if the outputs of MARKAL were fed back to the CGE model, it would be impossible in practice to form a consistent model after repeated iteration. As for the CGE models themselves, the two models used different types of functions, chose different parameters, simulated under different assumptions, and hence obtained different results.

The results of Jorgenson et al. (1998) were rather optimistic. They showed that by charging a carbon tax of 9 Yuan/tc, and decreasing other enterprise taxation so that government revenues remained constant, CO_2 emissions could be reduced by 5 per cent.[1] Meanwhile, although economic growth might be slow during the initial period of carbon taxation, investment might subsequently be increased because enterprises would have more surpluses to invest after reduction of other taxes. Economic growth might be more rapid than in cases without carbon taxation. Moreover, the long-run economic growth might compensate for the short-run decline. Therefore, they argued that the Chinese government faced opportunities of 'double dividends'. That is to say, both long-run economic growth and environment amelioration could be realized at the same time through taxation reform.

Zhang's study shared a more pessimistic view in that a carbon tax of 205 Yuan/tc was necessary to reduce CO_2 discharge by 20 per cent while GDP declined by 1.52 per cent by the year 2010 as compared with the situation without carbon taxation. Moreover, GDP growth was always lower than the situation without carbon tax, given the assumption that other indirect taxes were reduced by 5 per cent and 10 per cent to compensate for the negative effects of carbon tax. Hence, in contrast to Jorgenson et al. (1998), Zhang's results showed a trade-off between economic growth and environmental amelioration.

Our study uses the CGE model of the Chinese economy, PRCGEM, as developed jointly by the Institute of Quantitative and Technological Economics, at the Chinese Academy of Social Sciences, and the Center of Policy Studies, at the Monash University, Australia. Costs of CO_2 emissions reduction through carbon taxation under different situations are analysed. It is a comparative static analysis because we consider only the effects of carbon taxation, not their paths of realization. An advantage of static analysis is that it deals with the short-run effects, in particular under the rigidity of factor markets. Our results show that impacts of carbon taxation on the whole economy will be much more moderate than Zhang's results show, but they are not as optimistic as the case of 'double dividends' in Jorgenson et al. (1998).

3 THE MODEL AND DATA

The original PRCGEM is a computable general equilibrium model describing China's economic activities. This section describes the relationships between economic activities and CO_2 emissions, the design modes of carbon taxation, and how to integrate carbon taxation into the model framework.

3.1 The Model

PRCGEM is a typical CGE model based on neo-classical general equilibrium theory. Its structure is similar to that discussed in Dervis et al. (1982), and Horridge et al. (1993). The basic assumptions of PRCGEM include the following:

1. Assumptions about producers:

- Each sector produces one kind of commodity;
- Price takers;
- Cost minimization;

- Nested Leontief/CES production function with substitution between domestic produced and imported inputs (that is, the Armington assumption) as well as substitution among labor, capital and land;

2. Assumptions about investors:

- Price takers;
- Cost minimization;

3. Assumptions about consumers:

- One category of consumers only;
- Utility maximization with budget constraints;
- Price receivers;
- Structure of consumption described by inlaid LES/CES function;
- Total consumption kept constant or as a fixed proportion of GDP;

4. Assumptions about exports:

- Incomplete substitution between products for export and those for domestic consumption;
- Demand curves for exports negatively inclined;

5. Assumptions about the government:

- Cost minimization;
- Total expenditures fixed or as a certain proportion of total consumption;

6. Assumptions about prices:

- Price equal to cost, that is, the zero profit assumption;
- Model is money-neutral; hence a price is required as the numeraire. In general, exchange rate is chosen as the numeraire;

7. Assumptions about commodity markets:

- Commodity markets always in equilibrium, market-clearing assumption.

In PRCGEM, all energy sources, including coal, oil, natural gas and electric power, and factors of production in the sectors are produced according to the Leontief production function. The drawback of its constant coefficient matrix is that no substitution among various energy sources is allowed. However, the substitution is important for our problem. Hence, we integrate the energy sources and factors of production with the CES function, with coal, oil, natural gas and electric power combined as energy according to the Cobb–Douglas function. The elasticity of substitution between energy and factors of production are assumed to be equal to one among various energy sources. Thus, the original inlaid PRCGEM production function is further inlaid in order to allow for substitution between energy and factors, and between energy sources.

3.2 Introduction of Carbon Taxation and CO_2 Emission Equations

For carbon taxation we calculate the carbon contents of three types of fossil fuel. In particular, we have:

$$C1(i,j) = A_j X1(i,j) \tag{7.1}$$

In the above formula, 1 means intermediate inputs; i is the sector; j is the type of fossil fuel; $C1(i,j)$ is the carbon content of fuel j used by sector i, $X1(i,j)$ is the quantity of fuel j used by sector i in tons or kilometers; and A_j is the carbon content of one unit of fuel j.

The above formula represents carbon consumed in the production process. What follows is that expended during the consumption process:

$$C3(j) = A_j X3(j), \tag{7.2}$$

where 3 represents household consumption, $X3(j)$ is the fuel j consumed by inhabitants, $C3(j)$ expresses the carbon content of fuel j consumed by inhabitants. Then we add up the above two items to arrive at the total carbon content of fuel j of domestic use (including both production and consumption use), that is,

$$C(j) = \sum_i C1(i,j) + C3(j) \tag{7.3}$$

Based on the total carbon content of fuel j, together with a transformation factor and a combustion efficiency factor, we could roughly calculate the CO_2 emission caused by fuel j usage, that is:

$$CO_2(j) = \alpha\lambda_j C(j) \tag{7.4}$$

in which the transformation factor, α, is equal to 3.67,[2] while λ_j is the combustion efficiency factor. Adding up all the CO_2 produced by fossil fuels, we arrive at the total CO_2 produced by fossil fuels in both production and consumption, that is:

$$CO_2tot = \sum_j CO_2(j) \tag{7.5}$$

It is relatively simple to introduce carbon taxation to the model. The purchasers' price of each kind of fossil fuel is equal to its producers' price plus carbon tax and other indirect taxes. The carbon taxes paid by purchasers are determined by carbon content of fossil fuels consumed and the carbon tax. In particular, the purchasers' price for producers is:

$$P1(i,j) = P0(j) + ITX(j) + A_j TC \tag{7.6}$$

where $P1(i,j)$ is the purchasers' price paid by sector i for fuel j, $P0(j)$ is the producers' price of fossil fuel j, $ITX(j)$ is the indirect taxes of fuel j, TC is the carbon tax. The carbon tax paid by sector i for fuel j is:

$$RTC1(i,j) = TC \times C1(i,j) \tag{7.7}$$

Similarly, the purchasers' prices of fossil fuels for consumers are:

$$P3(j) = P0(j) + ITX(j) + TC \tag{7.8}$$

Therefore, the carbon tax paid by consumers for the use of fossil fuels j is:

$$RTC3(j) = TC \times C3(j) \tag{7.9}$$

By adding up (7.7) and (7.9), we arrive at the total carbon tax revenue:

$$RTCTOT = \sum_j \sum_i RTC1(i,j) + \sum_j RTC3(j) \tag{7.10}$$

Finally, we add carbon tax of (7.10) into the fiscal revenue as defined in PRCGEM, and revise the GDP of the original model.

3.3 Data

The original PRCGEM consisted of a model of 118 sectors and another model of 33 sectors,[3] both being based on an input–output table of 1992. Our model consists of 34 sectors as listed in Table 7.A.

In order to calibrate the model, it is necessary to set several important elasticities. As no appropriate and detailed literature of elasticity evaluation is available, we can only refer to the data of other countries in setting the required parameters. The resulting elasticities are listed in Table 7.1.

In addition, we must know the carbon content of individual fossil fuel, its combustion ratio, loss of inefficiency, and other adjustment factors. This information is listed in Table 7.2.

Table 7.1 Elasticities used in PRCGEM

Main Elasticity	Value
Armington Elasticity of Substitution	2.0
Elasticity of Capital–Labor Substitution	See Table 7.B[4]
Elasticity of export demand	−5
CET Transform Elasticity	3–10

Source: Author estimated, referring to the data of other countries, such as Indonesia and Thailand.

Table 7.2 Carbon content, combustion ratio and inefficiency loss

Type of Fuel	Carbon Content[a]	Combustion Ratio[b]	Inefficiency Loss[b]
Coal	0.54tC/t	98.3%	1%
Oil	0.84tC/t	98.3%	1%
Natural Gas	$0.0006tC/m^3$	98.3%	1%

Notes:
a IPCC(1992)
b The World Bank(1994)

Table 7.3 World Bank estimate and our estimate of total CO_2 emission

	World Bank's (1994) for 1990	Ours (1992)
Total carbon	667.64Tg*	734.32Tg*
Total CO_2 emission	2380.84Tg*	2620.28Tg*

Note: * 1Tg= 1 million tons

Using equation (7.4) and the above data, we could estimate the total CO_2 emission in 1992 as shown in Table 7.3. There are some differences between the World Bank's estimate and ours. They refer to 1990, while we refer to 1992. We consider all carbon, while the World Bank only dealt with the carbon which became CO_2. In practice, some carbon was made into other products such as plastics and did not become CO_2. Unfortunately, we cannot derive data for concrete products from input–output tables. We just assume that the part of carbon which did not become CO_2 was due to low efficiency of combustion. So our estimate of CO_2 emission may be overstated[5] to some extent.

4 SIMULATION RESULTS

We have designed three scenarios to analyse their corresponding emission reducing costs: CO_2 emission reduction by 5 per cent, 10 per cent and 20 per cent. We also consider the scenario where other taxes are reduced after carbon tax is levied. Meantime, we distinguish between short-run and long-run costs of emission reduction. Therefore, we have carried out 12 simulations.

The so-called short run and long run do not represent a certain length of time; instead, they indicate the reaction process of macro-variables to changes in exogenous variables.[6] In the short run, we assume that aggregate consumption, investment and government expenditure remain constant; in addition, international revenue and expenditure are endogenous. The labor force can move freely among sectors, which means that there is non-voluntary unemployment, but capital cannot be mobile among sectors because capital in each sector has a specific usage. In fact, we assume that following an exogenous shock, firms adjust employment first and capital stock next, which means that the rate of return of capital could be different among sectors. In the long run, we assume that international revenue and expenditure remain balanced. Labor and capital are perfectly mobile and fully employed so that capital return rates are the same among sectors. This is the situation of full reaction to exogenous shock.

The simulation results can be divided into macro-effects, sectoral effects and regional effects. Tables 7.4 and 7.5 show, respectively, the short-run and long-run costs of emission abatement under different scenarios. In Tables 7.4 and 7.5, columns 2, 3 and 4 show the effects of CO_2 reduction by 5 per cent, 10 per cent and 20 per cent, respectively. Rows 2 and 3 show two scenarios. The first scenario is carbon tax without reduction of other taxes. In this case, the government revenue will increase. If government expenditure remains unchanged, government savings will increase. If total investment does not increase, the savings and investment of households will be squeezed out. The second scenario is a combination of carbon tax and reduction of enterprise taxes, keeping government revenue unchanged. It will counteract the aforementioned crowding-out effect.

Our results indicate that the cost increases with the increase of carbon taxation and amount of CO_2 reduced in the short term. For example, in Table 7.4 under the condition of carbon taxation without other taxation reduction, 5 per cent, 10 per cent and 20 per cent of CO_2 reduction will lead to 0.22 per cent, 0.47 per cent and 1.06 per cent of GDP decline respectively. When CO_2 reduction doubles, the decline in GDP will more than double.

If other taxes are reduced simultaneously with carbon tax, the decline of GDP becomes less and carbon tax will increase to some extent. For example, in order to reduce CO_2 by 20 per cent, real GDP declines by 1.06 per cent; while with reduction of other taxes, real GDP declines only by 0.34 per cent. The amount of taxation in the latter case is a little higher than that in the former case, which reflects that lowering tax will stimulate output growth and CO_2 emission, so more tax is needed to reduce these amounts. It is not the case of 'double-dividends' as concluded by Jorgenson et al. (1998).

Next, our result shows that the short-term cost of CO_2 reduction is greater than the long-term one. For example, under the condition of carbon taxation without other taxation reduction, 5 per cent, 10 per cent and 20 per cent of CO_2 reduction will lead to 0.05 per cent, 0.12 per cent and 0.34 per cent of GDP decline respectively in the short run (Table 7.4), as compared to 0.014 per cent, 0.01 per cent and 0.06 per cent decline in the long run (Table 7.5).

Third, we find that in the long run the cost increases first, then declines with the increase of CO_2 reduction. In Table 7.5, we can see that under the condition of carbon taxation with reduction of other taxation, 5 per cent of CO_2 reduction leads to a 0.014 per cent decline of GDP, but 10 per cent of CO_2 reduction only leads to a 0.01 per cent decline of GDP. It shows that higher carbon taxation does not necessarily mean a greater decline in GDP. The positive effect of other tax reduction can offset the negative effect of carbon tax on GDP. However, in the case of a 20 per cent CO_2 reduction, GDP declines 0.06 per cent, which is larger than in the case of a 10 per cent CO_2 reduction. It seems that there exists an optimal level of CO_2 reduction or taxation. The relation between CO_2 reduction and its effect can be seen in Figure 7.1.

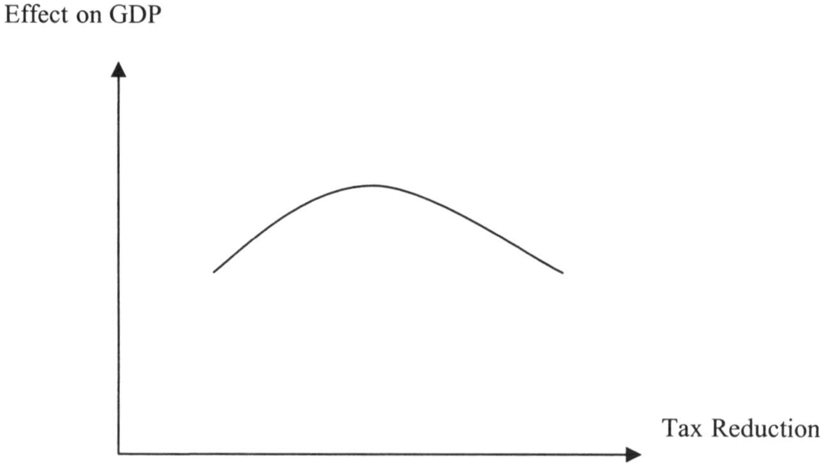

Figure 7.1 Effect of taxation reduction on GDP

Table 7.4 Short-run macro-impacts of CO_2 emission reduction under various situations

		CO_2 reduced by 5%	CO_2 reduced by 10%	CO_2 reduced by 20%
With carbon tax, no reduction of other taxes.	Real GDP	−0.22%	−0.47%	−1.06%
	Carbon Tax	13.75Yuan/Tc	29.13Yuan/Tc	66.11Yuan/Tc
With carbon tax, reduced enterprise taxes, government revenue neutral	Real GDP	−0.05%	−0.12%	−0.34%
	Carbon Tax	14.68Yuan/Tc	31.24Yuan/Tc	71.69Yuan/Tc

Table 7.5 Long-run macro-impacts of CO_2 emission reduction under various situations

		CO_2 reduced by 5%	CO_2 reduced by 10%	CO_2 reduced by 20%
With carbon tax, no reduction of other taxes	Real GDP	−0.06%	−0.13%	−0.36%
	Carbon tax	13.23Yuan/Tc	28.21Yuan/Tc	64.91Yuan/Tc
With carbon tax, reduction of other taxes, Government revenue unchanged	Real GDP	−0.014%	−0.01%	−0.06%
	Carbon tax	13.54Yuan/Tc	28.96Yuan/Tc	67.09Yuan/Tc

Furthermore, we decompose the effect of CO_2 emission reduction into output effect and structural effect according to equations (7.11) and (7.12).

$$CO_2tot = \frac{CO_2tot}{GDP} \times GDP \tag{7.11}$$

where CO_2tot represents amount of CO_2 emission reduction and GDP is in real value terms. By taking the differential logarithm of both sides of (7.11), we have:

$$co_2tot = s + gdp \tag{7.12}$$

(structural effects) (output effects)

where co_2tot and gdp represent the change rate of CO_2tot and GDP respectively, and $s = dLn(CO_2tot/GDP)$. From Table 7.6, we can see that the structural effects are much greater than the output effects.

Table 7.6 Decomposition of the CO_2 reduction effects

		Results of Decomposition		
		Output Effects	Structure Effects	Total
Short Run	No Reduction of Other Taxes	−0.47%	−9.53%	−10%
	Reduction of Other Taxes	−0.12%	−9.88%	−10%
Long Run	No Reduction of Other Taxes	−0.13%	−9.87%	−10%
	Reduction of Other Taxes	−0.015%	−9.985%	−10%

Finally, because carbon tax is based on the carbon content of fossil fuels, it is an *ad quantum* tax. However, we can convert it into an *ad valorem* tax. From (7.6) or (7.8), we have:

$$P(j) = P0(j)\left(1 + t_i(j)\right) + A_j \times TC \tag{7.13}$$

which states that the selling prices of fossil fuels are equal to producer's price plus indirect taxes and carbon tax per unit of fuel. Obviously,

$$P(j) = P0(j)\left(1 + t_i(j) + \frac{A_j \times TC}{P0(j)}\right) \tag{7.14}$$

where $\left(A_j \times TC\right)/P0(j)$ is the *ad valorem* tax of fuel j, which can be expressed as $TCE(j)$.

By differentiating the logarithm of $(1+TCE(j))$, we arrive at the percentage variation of the *ad quantum* tax corresponding to carbon tax variation.

$$d\ln\left(1 + TCE(j)\right) \times 100 = \frac{A_j dTC}{P0(j)} \times 100 \tag{7.15}$$

By using Equation (7.15), we can transfer carbon tax in Table 7.4 and Table 7.5 into corresponding energy tax. For example, 5 per cent of CO_2 reduction without other tax reduction in the short term needs carbon taxation of 13.75 Yuan/tc. From Table 7.2, carbon content per ton of coal is 0.54t. From Table 7.7, the price of coal is 57.02 Yuan/t. So 13.75 Yuan/tc of carbon tax is equivalent to 7.4 Yuan/t of energy tax (that is, 13 per cent of price).

Table 7.7 Consumption of fossil fuels, 1992, quantity and value

Type of Fossil Fuel	Physical consumption in 1992[a]	Value of Fossil Fuel[b]	Price of Fossil Fuel[c]
Coal	1.14 billion ton	65 billion Yuan	57.02Yuan/ton
Oil	0.13 billion ton	60.9 billion Yuan	468.46Yuan/ton
Gas	15.8 billion M^3	1.37 billion Yuan	0.09Yuan/M^3

Notes:

a China State Statistics Bureau (1994).

b China State Statistics Bureau (1996).

c Calculated from columns 2 and 3.

5 SECTORAL AND REGIONAL IMPACTS

Like the aforementioned macro results, there are different sectoral impacts with different scenario designs. The greater the CO_2 reductions, the greater are the sectoral impacts on GDP. For the sake of simplicity, we will only discuss the case of 10 per cent CO_2 reductions.

The key to understanding the sectoral impacts is to determine the sectoral carbon intensity. A sector with high carbon intensity will suffer more from carbon taxation and vice versa. In Table 7.8, we use two ways to measure carbon intensity: (1) share of fossil fuels cost in total cost, and (2) tons of carbon consumed per one Yuan of output. We list five sectors with maximum carbon intensity and five sectors with minimum carbon intensity. We can see that in either case the five sectors with maximum and minimum carbon intensity are the same, only appearing in a different order.

In Table 7.9 and 7.10, we list five sectors with maximal decline and five sectors with minimal decline of outputs and show the size of these changes. For example, from Table 7.9 under the condition of carbon taxation without other taxation reduction, output of the coal mining sector will decline the most, that is by 10.68 per cent relative to the condition of no carbon taxation.

Considering Table 7.9 and 7.10 together, we notice that the five sectors with the greater decline in output basically are the five sectors with maximal carbon intensity in Table 7.8. The sectors of coal mining, coke and natural gas always belong to the five sectors with maximal decline of output in both tables. However, for the five sectors with minimal decline of output, this kind of relationship no longer exists. In the short term, only the construction sector belongs to the sectors of minimal decline of output.

It is notable that in the short term, output of the textile and garment sector will decline the most. Although carbon tax cannot directly increase the cost of intermediate input in the textile and garment sector, it leads to an increase in consumer prices and wages because both sectors are labor-intensive.

In addition, the ratios of export to output are quite high in both sectors. Therefore, the price increase leads to the decline of their exports and outputs. However, in the long term, they are the sectors with high growth potentials because of the decline of real wages and inflow of capital. Obviously, there are big differences between the short-term and long-term effects.

Sectoral analysis shows more clearly that the other tax reduction can counteract the effect of carbon taxation. For example, in Table 7.9, without other tax reduction, carbon taxation leads to decline of output in every sector; but when reducing other enterprise tax, output appears to increase to some extent in food processing, electric power, restaurants, and so on, but not in coal mining.

Finally, we notice that in the long term, the electric power sector will have a larger growth following carbon taxation. The reason is that we assume the substitution exists between different kinds of energies. Due to the taxation on fossil fuels, people will use electric power to replace the direct use of fossil energy. According to the assumptions of our model, a 1 per cent increase in electric power and a 1 per cent decline in coal will keep the energy supply unchanged. However, the CO_2 emission following a 1 per cent increase in electric power will be less than the CO_2 emission following a 1 per cent increase in coal.

Table 7.8 Carbon intensity measured in two ways

Measured in value				Measured in real terms			
5 sectors with max. intensity		5 sectors with min. intensity		5 sectors with max. intensity		5 sectors with min. intensity	
Sector	Intensity	Sector	Intensity	Sector	Intensity, TC/Yuan	Sector	Intensity, TC/Yuan
Oil Refinery	52%	Construction	0.06%	Coke	0.0033	Construction	5.49E-06
Coke	38%	Farming	0.08%	Electric Power	0.0014	Electronic	7.46E-06
Electric Power	18%	Electric	0.11%	Oil Refinery	0.00094	Farming	7.57E-06
Coal	3.5%	Apparel	0.13%	Coal	0.00033	Electric	9.07E-06
Building Materials	3%	Electronic	0.16%	Building Materials	0.00026	Apparel	1.18E-05

Table 7.9 Short-run sectoral impacts

With carbon taxation, no reduction of other taxes				With carbon taxation and reduction of other taxes, fiscal revenue unchanged			
5 sectors with maximum impacts on output		5 sectors with minimum impacts on output		5 sectors with maximum impacts on output		5 sectors with minimum impacts on output	
Sector	Impacts on output	Sector	Impacts on output	Sector	Impacts on output	Sector	Impacts on output
Coal	−10.68%	Administration	0	Coal	−10.63%	Food	0.24%
Natural gas	−6.23%	Construction	−0.014%	Natural gas	−6%	Electric power	0.12%
Coke	−2.29%	Restaurants	−0.02%	Coke	−2.1%	Farming	0.02
Textile	−1.37%	Cultural and education	−0.15%	Oil Refinery	−0.86%	Restaurants	0.018%
Apparel	−1.33%	Repair	−0.16%	Minerals	−0.74%	Electronic	0.013%

Table 7.10 Long-run sectoral impacts

With carbon tax, no reduction of other taxes				With carbon tax and reduction of other taxes, fiscal revenue unchanged			
5 sectors with max. decline of output		5 sectors with increased output		5 sectors with max. decline of output		5 sectors with increased output	
Sector	Impacts on output	Sector	Impacts on output	Sector	Impacts on output	Sector	Impacts on output
Coal	−10.68%	Apparel	0.52%	Coal	−10.78%	Electric Power	1.03%
Natural gas	−6.75%	Electric power	0.44%	Natural gas	−6.76%	Apparel	0.45%
Coke	−2.0%	Textile	0.18%	Coke	−1.9%	Foods	0.4%
Oil refinery	−0.85%	Farming	0.15%	Oil refinery	−0.62%	Electronics	0.25%
Metallurgy	−0.74%	Foods	0.01%	Metal ores	−0.52%	Textile	0.16%

Table 7.11 Short-run regional impacts

With carbon taxation, and no reduction of ther taxes				With carbon taxation, and reduction of other taxes, fiscal revenue unchanged			
5 regions with max. decline of output		5 regions with least decline of output		5 regions with max. decline of output		5 regions with least decline of output	
Regions	Intensity	Regions	Intensity	Regions	Intensity	Regions	Intensity
Shanxi	−2.5%	Xizang	0	Shanxi	−2.27%	Xizang	0.1%
Ningxia	−1.1%	Hainan	−0.23%	Ningxia	−0.84%	Hainan	0.01%
Heilongjiang	−0.94%	Qinghai	−0.42%	Heilongjiang	−0.66%	Guangdong	−0.1%
Neimenggu	−0.78%	Guangxi	−0.43%	Neimenggu	−0.5%	Fujian	−0.12%
Liaoning	−0.76%	Yunnan	−0.45%	Henan	−0.43%	Zhejiang	−0.123%

Table 7.12 Long-run regional impacts

Regions with carbon taxation, max. decline of output				Regions with carbon taxation, reduction of other taxes, fiscal revenue unchanged			
Shanxi	−2.32%	Xizang	0.05%	Shanxi	−2.19%	Xizang	0.16%
Ningxia	−0.87%	Hainaan	0.006%	Ningxia	−0.72%	Hainan	0.12%
Heilongjiang	−0.68%	Zhejiang	−0.07%	Heilongjiang	−0.53%	Guangdong	0.06%
Neimenggu	−0.51%	Fujian	−0.087%	Neimenggu	−0.37%	Zhejiang	0.05%
Liaoning	−0.45%	Guangdong	−0.088%	Henan	−0.31%	Fujian	0.04%

As for the regional results, we transfer the impacts of carbon taxation on sectors into regions according to the sectors' output shares in each region, under the assumption that the same sector's carbon taxation impact makes no difference among regions. These results are listed in Tables 7.11 and 7.12. If those sectors with a greater decline in output have a larger share of output in the region, the output of the region will decline. It is therefore not difficult to understand why Shanxi, Ningxia, Heilongjiang and Inner Mongolia are always the provinces where outputs decline the most; while Hainan, Guangdong, Zhejiang and Fujian are always the provinces with the greatest increase in output.

6 CONCLUSIONS AND PROSPECTS

We have used a computable general equilibrium model to analyse the possible costs for controlling CO_2 emission in China with carbon taxation. The costs depend on several factors such as the required extent of emission reduction, short-run or long-run reaction to carbon tax, and whether other taxes are reduced. Our results indicate that impacts on the whole economy will be much more moderate than impacts on individual sectors. Simulation has shown that under the worst conditions (20 per cent reduction of emissions in the short run, without reduction of other taxes), the real GDP will decrease by 0.96 per cent. Under the best conditions (5 per cent reduction of emissions in the long run, with reduction of other taxes), the real GDP will decline by only 0.016 per cent.

Our sectoral analysis shows that outputs of coal, oil refinery and coke industry will decline the most, while sectors using less energy, such as clothing and food, increase their outputs slightly. It is noteworthy that output of electric power will increase because of the substitution of electricity for fossil energy. In the regional analysis, Shanxi, Ningxia, Heilongjiang, and Neimenggu are regions where outputs decrease the most, while Hainan, Guangdong, Zhejiang and Fujian will mostly expand in output.

As evidenced by our analysis, because of the substitution relationship in the economy, it is not necessary for GDP to decline to the same extent as the decrease in energy production in response to carbon taxation. From this point of view, it is probably not a calamity to levy a carbon tax. On the other hand, substitution in the economy means changes in the structure of production and consumption. We assume no adjustment costs in the model, which is unrealistic in practice. Workers in coalfields cannot simply move to other occupations without retraining. People must buy electrical equipment before they can replace coal with electricity. This type of adjustment cost will make a carbon tax unattractive and therefore it will be necessary to reduce other taxes at the same time.[7]

APPENDIX

Table 7.A Classification of sectors in the model

Code	Sector	Code	Sector	Code	Sector
1	Farming	13	Oil Refinery	25	Other Industries
2	Coal Mines	14	Coke	26	Construction
3	Oil Wells	15	Chemicals	27	Transportation
4	Natural Gas	16	Building Mat.	28	Trade
5	Mineral Mines	17	Metallurgy	29	Restaurants
6	Other Metallic Minerals	18	Metallic Products	30	Passenger Transport
7	Food	19	Machinery	31	Public Utility
8	Textiles	20	Transportation Equipment	32	Culture and Education
9	Sewing	21	Electric Machinery	33	Finance
10	Furniture	22	Electronic	34	Administration
11	Papermaking	23	Weights and Measures		
12	Electric Power	24	Repair		

Table 7.B Capital–labor substitution coefficients of individual sectors

CODE		CODE	
1	0.5	18	1
2	0.77	19	1
3	1.79	20	0.5
4	1.79	21	0.5
5	0.62	22	0.9
6	0.8	23	0.9
7	0.8	24	0.8
8	1.07	25	0.8
9	0.8	26	0.5
10	1	27	0.5
11	1	28	0.5
12	0.8	29	0.5
13	1.59	30	0.5
14	1.59	31	0.5
15	0.9	32	0.5
16	1	33	0.5
17	0.8	34	0.5

NOTES

1. Relative to cases without carbon taxation. All conclusions hereafter refer to comparison between cases with carbon taxation and without carbon taxation.
2. That is, the molecular weight of CO_2, 44, divided by the molecular weight of carbon, 12.
3. The two models have an identical structure, with different degrees of aggregation.
4. Estimated by Professors Gong Yi and Loyed.
5. If the World Bank estimate is accurate, and if the emission reached what we estimate for 1992, the annual average rate of growth of CO_2 emission will be an incredible 5.0 per cent.
6. Thus, we don't suppose that changes in capital stock are caused by investment, increase of labor force and technical progress in the long run.
7. Our analysis of short-run costs could be seen as increasing certain adjustment costs, for example, capital cannot move freely among sectors. Obviously, the short-run costs are much greater than the long-run costs.

REFERENCES

China State Statistics Bureau (1994), *The Chinese Statistical Yearbook*.

China State Statistics Bureau (1996), *The 1992 Input–Output Table of China*.

Dervis, K., D.J. Melo and S. Robinson (1982), *General Equilibrium Model for Developing Policy: A World Bank Research Publication,* Cambridge University Press, pp. 131–248.

Horridge, J.M., B.R. Parmenter and K.R. Pearson (1993), 'ORANI-F: A general equilibrium model of the Australian Economy', *Economic and Financial Computing*, **3** (2), London.

IPCC (1992), *The Intergovernmental Panel on Climate Change 1990 and 1992 Assessment*, WHO/UNE.

Jorgenson, D., M.S. Ho and D.H. Perkins (1998), 'China's Economic Growth and Carbon Emissions', in M.B. McElroy, C.P. Nielsen and P. Lydon (eds), *Energizing China*, Cambridge: Harvard University Press, pp. 301–42.

The World Bank (1994), 'China: Issues and Options in Greenhouse Gas Emission Control', Working paper or project report, No. 4.

Zhang, Z.X. (1996), 'Integrated Economy–Environment Policy Analysis: A Case Study for the People's Republic of China', Ph.D. thesis, Department of Economics, University of Wageningen.

8. Cost of reducing CO_2 emissions in Japan

Masahiro Kuroda and Koji Nomura

1 INTRODUCTION

The main purpose of this chapter is to examine the possibility as well as the cost in Japan of achieving the target figure for greenhouse gas emissions established at the COP3 as a countermeasure against global warming in the year 2010. Various policies are being considered to reach this target figure. For example, the introduction of a carbon tax is one possibility.[1]

Based on an evaluation of the economic effect when the tax is introduced, it is predicted that there will be complicated repercussions that are reflected by the economic structural interdependence mechanism. In this chapter, the economic means of implementing environmental protection countermeasures are broadly divided into two groups: market-oriented instruments such as a carbon tax and emissions trading, and regulated instruments such as specific energy-saving measures and other government regulations. The evaluation of the effectiveness of the policy adopted should be judged synthetically in the light of the cost effectiveness of environmental protection, efficiency and equity.

Our chapter proceeds as follows. We adopt the framework of the multi-sectoral general equilibrium model for Japan as described in Chapter 4. In Section 2, we summarize the model used here, and then, by using this model, we conduct several policy simulations in Sections 3 and 4 that explore alternative mechanisms for achieving the target figure for the emission of global warming gases. We are specifically interested in the effect that these policies will have on the Japanese economy. Finally, in Section 5, we discuss issues concerning the system design of global warming measures including the use of market-oriented measures as well as other negotiated measures such as CDM and JI.

2 CARBON TAX IN A MODEL SYSTEM[2]

In this section, we begin with a description of the logical flow (Figure 8.1) of the impacts from a carbon tax system on the economy in our model.[3] The short-term effect on the supply side is given on the left-hand side and the effect on the demand side on the right. The long-term effect is given in the lower right corner in Figure 8.1. To begin with, taxing carbon causes energy input prices to rise. The tax is proportional to the amount of CO_2 emitted from energy combustion. The impact of this increase in energy input prices is concentrated in energy-intensive industries. The short-term supply curve for this sector thus shifts to the left. Since the various industrial sectors are dependent on each other through intermediate input transactions, the increase in energy input prices will ripple through the supply chain, directly resulting in an increase in energy prices and indirectly in an increase in raw material prices throughout all areas of industry, therefore pushing the supply curve even farther to the left.

Assuming that the demand curve maintains its position, the point of equilibrium between supply and demand will shift from A to B. The resulting value-added will change in the various areas of industry, which will lead to a change in incomes. The increase in primary energy prices will change demand. On the other hand, if we look at investment behavior, the change in relative prices that will come from higher energy prices will change industry's supply capacity and resource allocation, depending on the substitution among capital, labor, raw materials and alternative energy sources in each specific industry. Based on the relationship of capital to alternative energy sources, a relative increase in energy prices will most likely encourage investment in energy conservation and give rise to an increase in the demand for investment capital. If a decline in consumption demand and net exports offsets investment capital demand and overall demand falls, the change in the demand structure will shift the demand curve for this term to the left, except for some capital. As a result, the equilibrium point will shift from B to C as shown in Figure 8.1.

As is shown in the lower right corner of Figure 8.1, the intersection of the increase in capital stock and the increase in fixed costs is predicted to move upward from time t to time $t+1$. However, because of increased productivity that centers on greater energy efficiency, curvature changes turn the supply curve in $t+1$ to the right. As a result, because the balance point for $t+1$ and beyond changes from A to B, economic activity will pick up, leading to lower prices and increased production. The dynamic result is very meaningful from the perspective of balancing environmental protection and economic growth. Artificial price changes caused by carbon taxing will, in the short run, cause a general increase in prices and a fall in real GDP, but rising productivity centered around the improvement in energy efficiency improvement will, in the long run, bring economic growth and new competitive strength through the introduction of new technology.[4]

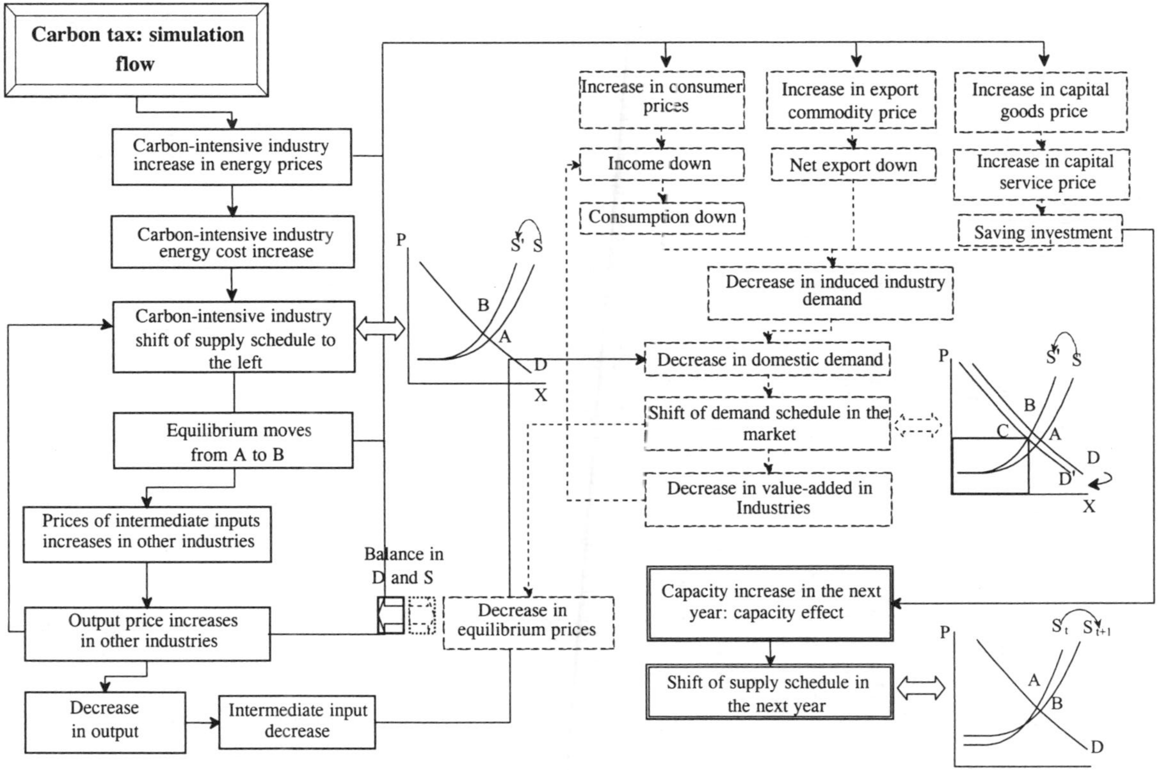

Figure 8.1 Logical flow of the simulation for carbon tax

3 MARGINAL COST OF CO_2 REDUCTION

Being a Baumol–Oates tax, amounts of the carbon tax can only be calculated based on the original setting of total emissions.[5] The marginal cost is obtained as the amount of the real carbon tax needed to attain that level, using our multi-sector general equilibrium model. Here, to enable comparison, the real carbon tax is converted to a 1985 price base with consideration being given to a general rise in prices (endogenously obtained) that would be caused by the carbon tax needed to achieve the targeted reductions.[6]

Figure 8.2 illustrates the marginal cost curves for CO_2 reduction focusing on the short-term effects based on our model's results. This figure depicts the marginal cost of the emission reduction in additional millions of tons of carbon emissions above the BaU case. The BaU scenario used in our model puts total CO_2 emissions for 2010 at 349 million t-C and requires a reduction of 51 million t-C in order to stabilize emissions at the 298 million t-C target level for 1990. An important assumption in this BaU scenario is that nuclear power plants will have a 70 million kW generation capacity by 2010. The top panel of Figure 8.2 plots the marginal cost curves for the years 2000 and 2010, whereas the resulting effects on real GDP and the GDP deflator are plotted in the bottom panel. The marginal cost of reducing CO_2 is no greater than the carbon tax rate.

According to Figure 8.2, the marginal costs increase only slightly so that the rise in costs is gradual to the point that the curve is almost linear. In 2000, a 965 yen (1985 price base) carbon tax would be needed to reduce CO_2 emissions by 1 million t-C, a 9997 yen carbon tax to reduce them by 10 million t-C and a 20822 yen carbon tax to reduce them by 20 million t-C. Approximately a 1000 yen carbon tax must be added to reduce CO_2 emissions by an additional 1 million t-C in 2000. Similarly, in 2010, an increase in the carbon tax of 700–800 yen will be needed to reduce CO_2 emissions by an additional 1 million t-C. This means that, for the same amount of reduction, taxes in 2010 are 70–80 per cent of year 2000 taxes, the reason for this being that taxes in each year completely depend on differences in the efficiency of the economic conditions, the level of the technology, and the amount of total emissions.

Let us also look at the effect that carbon taxing could have on the economy as a whole. For an additional reduction in emissions by 1 million t-C, real GDP in 2000 would fall by an additional 42–47 billion yen and the GDP deflator would rise by an additional 0.09–0.11 per cent. Similarly, in 2010, real GDP would fall by an additional 33–37 billion yen, and the GDP deflator would rise by an additional 0.05–0.07 per cent. For a 20 million t-C reduction, in the year 2000 there would be a 1.92 per cent rise in prices and a drop in real GDP of 1.024 trillion yen in the short run (without the reduction, prices would fall by 2.0 per cent without a change in real GDP), while in 2010, prices would rise by 1.17 per cent and GDP would decline by 7.275 billion yen (similarly a drop of about 1.1 per cent).

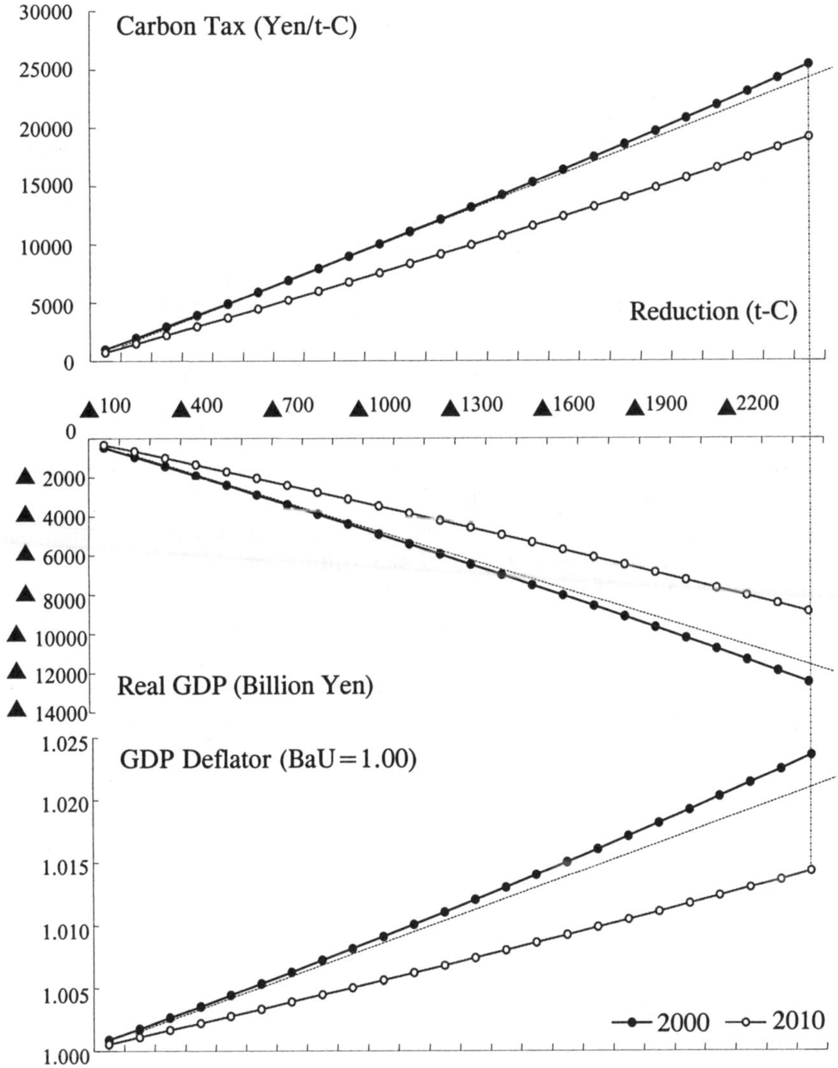

Figure 8.2 Marginal cost curve for CO_2 reduction and the short-term effects

4 POLICY SIMULATION

4.1 Simulation to Stabilize CO_2 Emissions at the 1990 Level

As was mentioned earlier, the government's policy scenario is that of stabilizing total CO_2 emissions in 2010 at the 287 million t-C level for 1990 by reducing 27 million t-C through increased use of nuclear power generation and 60 million t-C through energy conservation efforts. Results from the simulation are given in Table 8.1. Here it is assumed, as the means for stabilizing CO_2 emissions at 1990 levels, that a flat carbon tax is levied on the amount of CO_2 emitted through the consumption of fossil fuels. The carbon tax is proportional to the carbon content of each specific fossil fuel. The tax is levied only when fossil fuels are used, that is to say, only when CO_2 is directly emitted as a result of the combustion of fuel. Nevertheless, the same tax rate per unit of carbon is applied to industry, transportation and consumers.

In the BaU scenario, the 349 million t-C of CO_2 emissions in 2010 are reduced by 51 million t-C to 298 million t-C. In this case, targeted reduction cannot be attained solely by taxing carbon in the year 2010, so taxes are levied for two years from 2009. The amount of tax needed to attain the targeted reduction was endogenously obtained at 29711 yen per one ton of carbon for 2009 and 51509 yen per one ton of carbon for 2010. In this case, total carbon tax revenue comes to 9.483 trillion yen in 2009 and 15.265 trillion yen in 2010. As a result, real GDP falls by about 1.3 per cent in 2009 in the BaU case. The decrease in 2010 is 0.36 per cent from 669205 billion yen to 666739 billion yen. However, this GDP level for 2010 is given for the standard case, in which a carbon tax is not introduced. If the tax were introduced in 2009 and not in 2010, real GDP would fall by 1.42 per cent in 2010 in comparison with the BaU case.[7]

The introduction of a carbon tax will improve energy efficiency in all energy-using sectors and this will serve to increase GDP. Accordingly, if the tax were introduced in 2009, the improved energy efficiency that the tax would bring about would lower prices and increase GDP in 2010. Although further carbon taxing in 2010 will slow economic growth, part of that will be made up by the energy efficiency improvements of the previous year. These results show that carbon taxing in 2009 and 2010 will impose a burden upon consumers in the short run, but improved energy efficiency will increase supply capacity and ultimately offset that burden.

Table 8.1 Simulation results of stabilizing CO_2 emission at the 1990 level

	unit	1985	1990	1995	2000	2005	2008	2009	2010	95–00	00–05	05–10	98–10
Exogenous variables				Volume							growth rate (%)		
Population	10 thousand	12105	12361	12557	12689	12768	12777	12772	12763	0.21	0.12	−0.01	0.09
No. of households	10 thousand	3875	4084	4375	4666	4925	5060	5103	5145	1.29	1.08	0.88	1.04
Money supply	1 trillion yen	2952	4291	4609	5246	6224	7018	7304	7602	2.59	3.42	4	3.52
Exchange rate	yen/dollar	238.54	144.8	94.06	122.55	127.56	130.65	131.7	132.76	5.29	0.8	0.8	0.8
Nuclear capacity factor	10 thousand Kw	2452	3148	4119	4510	5435	6310	6655	7000	1.81	3.73	5.06	3.68
Nominal inventory investment	1 billion yen	2777	2165	3110	3010	3034	2989	2969	2955	−0.65	0.16	−0.52	0.14
Nominal social security benefit	1 billion yen	34918	48823	64597	82944	106502	123738	130082	136751	5	5	5	5
Nominal social security contribution	1 billion yen	26185	38957	50197	64454	82760	96154	101084	106266	5	5	5	5
Nominal governmental consumption expenditure	1 billion yen	30685	38807	47555	52556	58084	61675	62921	64192	2	2	2	2
Nominal governmental capital formation	1 billion yen	11269	15399	22549	24920	27541	29244	29835	30437	2	2	2	2
International trading	10 billion $	1936	3466	5147	5605	5892	6072	6133	6194	1.7	1	1	1
Endogenous variables				Volume							growth rate (%)		
Real GDP	1 billion yen	309716	455016	539683	531514	580948	630308	640887	666739	−0.31	1.78	2.75	1.72
GDP deflator	1985=1	1.026	0.871	0.764	0.914	0.95	0.945	0.955	0.94	3.58	0.79	−0.22	0.96
Interest rate		0.063	0.042	0.049	0.067	0.07	0.068	0.067	0.063	6.04	0.8	−2.04	0.32
CO_2 emission	million ton-C	247	298	343	322	331	341	321	298	−1.28	0.53	−2.07	−0.52
CO_2 emission per capita	ton-C	2.0405	2.4138	2.7353	2.5386	2.591	2.672	2.5158	2.3379	−1.49	0.41	−2.06	−0.61
Generated power	10 billion kwh	60	88	111	116	123	128	130	127	0.85	1.07	0.71	0.83
CO_2 emission per fuel consumption	g/cal	119.01	115.61	113.23	112.96	112.52	112.39	113.5	113.05	−0.05	−0.08	0.09	0.01
CO_2 intensity	g/yen	0.798	0.656	0.636	0.606	0.569	0.542	0.501	0.448	−0.98	−1.25	−4.82	−2.25
Energy intensity	Kcal/yen	6.701	5.672	5.621	5.365	5.061	4.819	4.417	3.959	−0.93	−1.17	−4.91	−2.25
Average capital service price	1985=1	0.985	0.753	0.651	0.87	0.923	0.921	0.919	0.894	5.79	1.18	−0.64	1.19
Average wage rate (age group)	1985=1	1	0.977	0.841	1.036	1.139	1.187	1.201	1.2	4.18	1.88	1.05	2.06
Average energy price	1985=1	1.025	0.853	0.663	0.729	0.747	0.763	0.929	1.027	1.89	0.5	6.36	2.71
Average raw material price	1985=1	1.02	0.842	0.709	0.889	0.956	0.976	0.998	0.996	4.52	1.46	0.82	1.7

Table 8.1 (Continued)

	unit	1985	1990	1995	2000	2005	2008	2009	2010	95–00	00–05	05–10	98–10
Endogenous variables					Volume						growth rate (%)		
Share of nuclear power	%	27.76	24.18	27.02	27.38	30.23	35.27	36.83	37.6				
Share of coal-fired power	%	6.32	8.71	13.67	16.83	18.22	12.69	11.13	10.33				
Share of gas-fired power	%	21.64	20.25	18.19	20.46	20.21	19.62	19.42	19.82				
Share of petroleum-fired power	%	18.48	23.61	22.22	15.63	11.98	11.56	11.32	10.51				
Share of hydraulic power	%	13.39	11.29	7.7	8.68	8.64	10.4	10.94	11.17				
Share of new energy	%	0	0	0.02	0.1	0.16	0.22	0.23	0.24				
Share of private power generation	%	12.42	11.96	11.19	10.93	10.56	10.24	10.13	10.35				
Carbon tax per t-C	yen	0	0	0	0	0	0	29711	51509				
Nominal household consumption	billion yen	184685	228938	249008	288123	328238	355326	360882	368948	2.92	2.61	2.34	2.58
Nominal household savings	billion yen	86425	102558	1096963	134938	155268	168464	171516	174179	4.14	2.81	2.3	2.86
Nominal fixed private capital formation	billion yen	80833	105015	52499	72173	84575	89733	97810	99677	6.37	3.17	3.29	4.27
Nominal export (FOB)	billion yen	45733	45257	68777	86950	102459	116345	120744	127597	4.69	3.28	4.39	3.23
Nominal import (CIF)	billion yen	35430	37155	30220	40365	49931	57513	61403	65084	5.79	4.25	5.3	4.34
Governmental tax revenue	billion yen	58619	80953	75035	88829	100653	108380	119478	126797	3.38	2.5	4.62	3.36
Capital income tax	billion yen	21048	30445	24090	28077	31904	34212	34678	35038	3.06	2.56	1.87	2.43
Labor income tax	billion yen	18469	26232	26334	31389	35730	38808	39447	40454	3.51	2.59	2.48	2.72
Property tax	billion yen	7684	8859	8843	10766	11913	12473	12857	12848	3.94	2.02	1.51	2.16
Indirect tax	billion yen	10108	13186	13770	15980	17938	19292	19292	19326	2.98	2.31	1.49	2.12
Customs and import duties	billion yen	1310	2230	1998	2616	3169	3595	3721	3866	5.39	3.83	3.98	3.62
Carbon tax	billion yen	0	0	0	0	0	0	9483	15265	0	0	0	0

4.2 Simulations for Unaccomplished Additional Nuclear Power

In the BaU case, nuclear power is assumed to expand from its current level of 44.92 million kW to 70 million kW in 2010. CO_2 emissions are expected to decrease to 27 million t-C based on this increased nuclear generation. However, it would be very difficult to construct a nuclear capacity of 70 million kW in the current situation. Therefore, we will simulate the case where electricity supply is not developed as assumed by BaU. We develop four additional cases labeled Sim 1A, Sim 1B, Sim 2A and Sim 2B as shown in Table 8.2. These cases explore alternative assumptions about nuclear power and LNG-thermal power capacity and utilization in the case of Japan.

Sim 1A and Sim 1B assume a nuclear power capacity of 44.92 million kW and utilization of 83 per cent (1997 levels) and Sim 2A and Sim 2B assume a capacity of 56.20 million kW and utilization of 83 per cent (scheduled 2006 levels). Sim 1A and Sim 2A assume LNG-thermal power capacity of 70.9 million kW and utilization of 40.2 per cent and Sim 1B and Sim 2B assume a capacity of 70.9 million kW and utilization of 47.0 per cent. Required labor inputs and capital stocks in an overall electricity sector vary with each case. We assume that overall labor inputs are 305700 for BaU, 280100 for Sim 1A and Sim 1B, and 294300 for Sim 2A and Sim 2B. In addition, capital stock levels in 1985 value terms are assumed to be 84.4 trillion yen for BaU, 77.9 trillion yen for Sim 1A and Sim 1B, and 80.1 trillion yen for Sim 2A and Sim 2B.

Table 8.2 Scenarios for the case of unaccomplished nuclear power

			Nuclear Power Capacity		
			70 MkW	44.92MkW	56.20MkW
		Utilization	78%	83%	83%
	64.50 MkW	37.7%	BaU		
LNG Capacity	70.90 MkW	40.2%		Sim1A	Sim2A
	70.90 MkW	47.0%		Sim1B	Sim2B

Note: All capacity and utilization performance is assumed for 2010.

The larger the share that nuclear power commands, the lower the electricity price becomes. This is because the energy material cost could be reduced by the increases in the number of nuclear power plants, while the fixed costs associated with their construction would be increased. As a result, the electricity price, after energy demand and supply adjustments, will rise by about 4 per cent in 2010 in contrast with the value assumed by BaU in the cases of Sim 1A and Sim 2A, and will increase by about 3 per cent in the cases of 2A and Sim 2B with a relatively larger share of nuclear power capacity.

Table 8.3 Summary table of carbon tax simulation results

Case	GDP in real terms			CO_2 emission		Carbon tax		Indirect tax	Change in GDP	
(unit)	(billion yen)			(millions t-C)		(yen/t-C)		(%)	(%)	
	2009	2010	2010 Base	2009	2010	2009	2010		2010 BaU	2010 Base
BaU	649603	669205		345	349					
Carbon tax (51 M t-C)	640887	666739	676405	321	298	29711	51509		−0.3	−1.4
Sim1A	652675	673069		364	369					
Carbon tax										
General		665127			349		15925		−1.1	
Electricity		NA								
Industry	652041	663915	674916	361	349	1320	23523		−1.3	−1.6
Indirect Tax		661860			349	–	–	78	−1.6	
Sim2B	651776	672567		349	354					
Carbon tax										
General		671020			349		3937		−0.2	
Electricity		669540			349		18723		−0.4	
Industry		668935			349		10480		−0.5	
Indirect tax		679213			349	–	–	15.9	−0.3	

Due to the increase in the electricity price, energy prices in all sectors become higher. Under these circumstances, substitution between energy and capital will be promoted in investment plans. For this reason, real investment aimed at energy saving will expand in all sectors while increasing supply capacity. This will result in a shift in the supply curve to the right from the BaU-based curve and a reduce in the supply price. Consequently, real GDP will rise by about 0.57 per cent.

Taking Sim 1A and Sim 2B as examples, we explore alternative options for introducing a carbon tax system. In addition to a general carbon tax case, we also: (a) impose a carbon tax only on the power industry, (b) impose a carbon tax only on industrial use (including the power industry), and (c) impose a general indirect tax not specifically related to energy. The differences in these policy options are summarized in Table 8.3.

In the case of Sim 1A, if the plan is only concerned with reducing the same amount of carbon as in BaU, the tax cost per t-C will be about 16000 yen. Real GDP will decrease by about 1.18 per cent as a result of this tax burden. However, under the assumptions of Sim 1A, the unmitigated carbon emissions for Japan will be 20 million t-C higher because less nuclear power is available. In order to meet the 1990 CO_2 target, Sim 1A must reduce its emissions by another 20 million t-C which will require a carbon tax of about 53000 yen per t-C.

If the carbon tax is calculated as an energy tax only in relation to the electricity sector in Sim 1A, solutions to economic equilibrium cannot be determined. The model implies that the introduction of the tax system only in relation to the power sector is not feasible in Sim 1A and cannot be considered a policy choice.

When the carbon tax system is applied only to industries, the tax is required in both 2009 and 2010. The tax rate corresponds to about 23500 yen in 2010 and becomes higher than that based on the introduction of a general carbon tax.

When a general indirect tax is used for Sim 1A, the tax rate must be increased by about 78 per cent. Considering the value of real GDP, the indirect tax is the most inefficient scenario for CO_2 mitigation. If you compare the losses for these last two taxes in terms of their impact on GDP reductions, these last two losses would reach −1.36 per cent and −1.67 per cent in comparison with −1.18 per cent in the case of a carbon tax.

In the case of Sim 2B, if the additional reduction of 20 million t-C is included, a general carbon tax of 3900 yen is needed. Real GDP will decrease by about 0.23 per cent. Imposing the carbon tax only on the electricity sector requires a tax of 19000 yen, and a tax on all industries would amount to about 10500 yen. These tax rates are 4.7 times and 2.6 times, respectively, higher than a general carbon tax. When the carbon reduction is achieved through an indirect tax, the tax will be increased by about 16 per cent. In this case, real GDP decreases by about 0.39 per cent. The indirect tax, in this case, is the next most efficient tax following the imposition of a general carbon tax.

Of the four tax system options, the general carbon tax is the most rational since it imposes the smallest cost on society. In comparison, all the other tax choices lead to small biases in market choices that increase costs. In the case of Sim 1A, the worst case is the indirect tax. This is because the increase in indirect taxes will broadly affect the economy. In Table 8.4, the impacts of the carbon tax systems on households are calculated. The table estimates the earnings required to compensate each household so that they maintain their existing utility under each policy. This model classifies households into six categories according to the age of the head of the household. Since the form and scale of energy use varies with each age category, there are different effects for each group.

Table 8.4 Effect of the Sim 1A and Sim 2B cases on the household economy

Compensated variation by age bracket of householder						
	Sim 1A			Sim 2B		
	Carbon tax			Carbon tax		
	General	Electricity	Industry	General	Electricity	Industry
Age of Householder						
15–24	11047	N.A.	13016	2643	3616	5639
25–34	8858	N.A.	10532	2116	2870	4551
35–44	14411	N.A.	17369	3447	4588	7511
45–54	19908	N.A.	24114	4759	6332	10431
55–64	10193	N.A.	12144	2432	3181	5245
65–	6055	N.A.	7479	1550	1994	3223
Energy consumption ratio to household economy						
Fuel consumption	0.8646	N.A.	0.9835	0.9698	0.9975	0.9929
Power consumption	0.9244	N.A.	0.8893	0.9860	0.9380	0.9632
Energy consumption	0.8791	N.A.	0.9605	0.9737	0.9829	0.9856

Notes:
1. Unit of compensated variation: Yen per month by household.
2. Energy consumption ratio: differences from BaU.

From observing each age category of householders, it is seen that policies have the greatest effect on the 45–54 age groups. In the case of Sim 1A, further reductions in CO_2 emissions are needed due to unaccomplished additional nuclear power. If the supply capacity of nuclear power stays at its current level (44.92 million kW), a household would need 20000 yen per month to compensate for the general carbon tax. Under similar circumstances, if the carbon tax were imposed only on industries, a household would need compensation of about 24000 yen. These costs are quite high, for the losses amount to more than 2–3 per cent of total family income each month. In Sim 2B where there is more nuclear power capacity, the tax has a smaller effect on households. The cost corresponds to about 4700 yen per month in the 45–54 age categories, and to about 1500 yen in the 65 and above age categories.

Each plan has implications for energy saving in the economy. In Sim 1A, the economy must reduce energy consumption by about 20 per cent as required in industries. In Sim 2B, the economy must only reduce energy consumption by about 3 per cent with a general carbon tax, and by 6 per cent with a carbon tax only on electricity.

5 POLICY SUGGESTIONS

Given the policy simulations for a carbon tax system, some system design problems should be considered carefully. First, the marginal cost of making substantial improvements in energy efficiency is high in Japan. Improving energy efficiency is thus expected to be expensive. Because the current level of energy efficiency in Japanese society is already high, additional improvements in Japan will be expensive. It is therefore likely that improving energy efficiency in other less-developed countries may be significantly less expensive than local expenditures. Japan should therefore actively step up efforts to construct an international framework for emission trading and explore other flexible mechanisms like JI and CDM.

Second, it is possible to reduce the national burden through the purchase of emissions in an international market. When the international price of emissions is lower than the domestic price, emissions can be purchased abroad. By setting a national carbon tax in accordance with some emission goals, it will be possible to use the revenues to purchase overseas emissions.

Third, our results indicate that it is impossible to stabilize the current level of emissions at the targeted 1990 level in a single year. Based on our model, even if a flat carbon system is introduced, the maximum reduction is about 30 million tons. In this case, the tax system must be in operation for more than two years in order to meet the goal of emission stabilization. When any measures are adopted for more than two years, the improvement in energy efficiency resulting from the system reduces the price the following year and expands actual economic activities. On the other hand, these brisk economic activities increase carbon emissions after the second year. As a result, the reduction in real GNP will level off, but a carbon tax or an emission price will be increased.

Fourth, the most efficient tax is a carbon tax imposed on all fossil fuel sources in proportion to CO_2 emissions. The flat general carbon tax is the most efficient tax. The system should apply to all industries and households. However, our analysis does not include the transaction costs and monitoring costs of such a program. These should be considered as well.

In cases where a carbon tax is imposed only on industries, the cost efficiency is reduced in contrast to the flat general carbon tax. In addition, in the event of the tax being imposed only on industries, a ceiling is set on the volume of the reduction in the short term, and the cost efficiency worsens. Imposing a carbon tax only on the electricity sector is even less efficient and may not even be feasible because it may not be possible for this sector alone to meet the targets. An indirect tax, that is a general energy tax (taxation proportional to energy instead of a carbon tax), is the most inefficient tax we are considering.

Finally, we find considerable differences in the cost burdens of industries and households. In order to allow for a more equitable distribution of the burdens, it is possible to incorporate subsidies for industries or household economies. In the event that industries and consumers are considered independently in stabilizing emissions, emissions reduction in a household economy will become more important in the future. However, if an emissions trading market is established, households may have difficulty interacting in such a market. Furthermore, monitoring households and small firms may also be very difficult. When only industries are allowed to participate in the trading market, the volume of emission reductions in the short term will be limited, and the gaps in terms of burdens among industries and household economies will increase.

NOTES

1. See OECD–IEA (1994) and The Government of Japan (1997).
2. See Burniaux and Oliviera-Martins (1992) and Cropper and Oates (1992).
3. See Jorgenson and Wilcoxen (1993), Kuroda et al. (1995) and Kuroda and Nomura (1998).
4. Briefly, the oil shock that began at the end of 1973 drove up the CIF price on crude oil imports from \$3.29 per barrel in 1973 to \$10.79 per barrel the following year. During that same period, the GDP deflator reached 20.8 per cent and, for the first time in the postwar era, Japan's economy recorded negative growth of −1.2 per cent as real GDP fell from 230.2 trillion yen in 1973 (1990 price-adjusted) to 227.4 trillion yen in 1974. In the following year, the GDP deflator was reduced to 7.2 per cent and real GDP growth recovered to reach 3.1 per cent.
5. See Pigou (1932) and Baumol and Oates (1988).
6. See Dean (1994).
7. The BaU case estimated the fall in real GDP as a result of carbon taxing as follows;

Year	t	$t+1$	$t+2$	$t+3$
BaU case	BaU→	BaU→	BaU→	BaU→
Carbon Taxing (a)	Sim→	BaU(Sim-a)→	BaU(Sim-a)→	
Carbon Taxing (b)	Sim→	BaU(Sim-b)→		

In addition to the BaU case, another case (BaU [Sim-a]) proposes carbon taxing in $t+1$ year but not in $t+2$ year. Technology and industrial structures at the start of $t+2$ year differ between the BaU case and carbon taxing (a). The description in this section first compares BaU in $t+2$ year against Sim, and then BaU (Sim-a) against Sim.

REFERENCES

Baumol, W.J. and W.E. Oates (1988), *The Theory of Environmental Policy*, second edn, Cambridge: Cambridge University Press.

Burniaux, J.M. and J. Oliviera-Martins (1992), 'The Effect of Existing Distortions in Energy Markets on the Cost of Policies to Reduce CO_2 Emissions: Evidence from GREEN', OECD Economic Studies, No. 19, Paris: OECD.

Cropper, M.L. and W.E. Oates (1992), 'Environmental Economics: A Survey', *Journal of Economic Literature*, **30** (2), pp. 675–740.

Dean, A. (1994), 'Costs of Cutting CO_2 Emissions: Evidence from Top Down Models', in OECD/IEA (1994).

The Government of Japan (1997), *Kyoto Protocol to the United Nations Framework Convention on Climate Change*.

Jorgenson, D.W. and P.J. Wilcoxen (1993), 'Reducing US Carbon Dioxide Emissions: The Cost of Alternative Instruments', *Journal of Policy Modeling*, **15** (5/6), pp. 491–520.

Kuroda, M. and K. Nomura (1998), 'The Construction of the Multi-Sectoral General Equilibrium Model of the Japanese Economy and the Environmental Protection Policy Simulation: (1) The Environmental Protection Policy Simulation and the Construction of the Multi-Sectoral General Equilibrium Model', KEIO Discussion Paper, No. 15, Keio University Institute for Economic and Industrial Studies.

Kuroda, M., I. Shimpo, K. Nomura and N. Kobayashi (1997), *KEIO Database: The Measurement of Production and Inputs of Capital and Labor*, KEIO Monograph Series, No. 8, Keio University Institute for Economic and Industrial Studies.

Kuroda, M., K. Nomura, N. Kobayashi, M. Kuninori, H. Kimiko and H. Tomita (1995), 'Reduction of Carbon Dioxide Emission and its Distributional Impacts', presented at the JDB Symposium on the Environment and Sustainable Development, Hakone, November.

OECD–IEA (1994), *The Economics of Climate Change: Proceedings of an OECD–IEA Conference*, Paris: Organization for Economic Cooperation and Development.

Pigou, L.C. (1932), *The Economics of Welfare*, 4th edn, London: Macmillan.

Shah, A. and B. Larsen (1992), 'Carbon Taxes, the Greenhouse Effect and Developing Countries', Working Paper WPS957, Policy Research Division, World Bank, Washington, DC.

9. Greenhouse gas mitigation through energy crops in the US with implications for Asian Pacific countries

Uwe A. Schneider and Bruce A. McCarl

1 INTRODUCTION

Energy crops have been explored in the US since 1978 (US DOE, 2001) about 20 years before the Kyoto Protocol was established. Major initial objectives involved reducing the dependency on foreign petroleum reserves and providing cleanly burning fuels. The potential to mitigate emissions of clean gases such as carbon dioxide was not emphasized until the 1990s when countries started to seriously negotiate greenhouse gas emission reduction programs.

Almost all energy crops produced in the US today are still subsidized. Ethanol suppliers receive, on average, a $0.54 per gallon subsidy, which is even greater than the 1998/1999 wholesale price of gasoline of $0.46 per gallon (Yacobucci and Womach, 2000). Governmental incentives to promote biomass power include project co-funding; various tax credits, deductions and exemptions, as well as direct subsidy payments (Badin and Kirschner, 1998).

Greenhouse gas emission mitigation efforts could improve the economics of energy crop production. If a market or governmental institution values carbon emission savings, energy crops would yield additional revenues equal to carbon price times net emission savings relative to an energy equivalent amount of fossil fuel. The question then becomes, what carbon price level is needed to make energy crops economically feasible? In answering this question one must analyse both energy crop possibilities and traditional agricultural production (Schneider, 2000) because:

1. Large-scale production of energy crops would substantially reduce the amount of land devoted to traditional food and fiber crops. As a consequence, aggregate food and fiber production would most likely fall,

causing commodity prices and land values to rise. The increased opportunity cost as reflected by higher land values may cause a strong negative feedback to large-scale efforts on promotion of energy crops.

2. General mitigation incentives will encourage a variety of agricultural strategies. McCarl and Schneider (1999 and 2000) grouped agricultural GHG mitigation options into three broad categories: (a) reductions in agricultural-based emissions, for example through diminished use of fossil fuels, fossil fuel intensive inputs, or livestock herd size reductions; (b) enlargements of agricultural-based sinks, for example through afforestation or tillage changes, and (c) increased production of commodities such as energy crops which offset emissions in other sectors of the economy.

Some agricultural mitigation strategies are mutually exclusive, some are complementary, and most interfere with traditional agricultural production. Hence, an independent analysis of a large-scale production of energy crops would most likely overestimate the economic potential.

2 THE US AGRICULTURAL SECTOR MODEL

For this analysis we used a new version of the US Agricultural Sector Model (ASM) (McCarl et al., 2001). ASM was first developed in the mid-1970s and has been used in many economic appraisals regarding environmental policies in the US (Chang et al., 1992). Schneider modified ASM to include GHG emission accounting and mitigation possibilities. This GHG augmented version is employed for this analysis and will hereafter be referred to as ASMGHG.

2.1 Scope of ASMGHG

ASMGHG depicts production in 63 US agricultural sub-regions endogenizing crop choice, irrigation choice, livestock numbers and livestock management. Commodity coverage is broad with more than 30 commodities considered including the major US feed and food grains, oilseeds, fiber, hay, silage, sweetener, cattle, sheep, poultry, dairy and hog commodities. Production of eight major internationally traded commodities in 28 rest-of-the-world regions is included with detailed international trade depiction for those commodities. Trade and consumption of over 50 other commodities are modeled at a more aggregate level. Production is gathered together into ten US marketing regions and in turn shipped on to processing, consumption or international markets.

ASMGHG solutions provide projections of land use and commodity production within the 63 US areas, commodity production in the rest of the world, international trade, crop and livestock commodity prices, processed commodity prices, agricultural commodity consumption, producer income effects, consumer welfare effects, and various environmental impacts from US agriculture and forestry.

2.2 Greenhouse Gas Features in ASMGHG

ASMGHG jointly incorporates all major GHG emission mitigation options available to agriculture (McCarl and Schneider, 2001). For this study we considered only potential strategies that are currently feasible. There are other strategies that may become practical in the near future as technology advances. However, we did not want to speculate as to when this might happen. Engineers are often overly optimistic about new technologies and fail to take economics into account. Currently included strategies are listed below. Schneider (2000) provides a detailed technical description of how these strategies are implemented in ASMGHG.

ASMGHG mitigation strategies through the agricultural sector:

1. Afforestation,
2. Production of energy crops for use in electrical power plants,
3. Production of ethanol to replace fossil fuel-based gasoline,
4. Soil carbon sequestration through crop, tillage, irrigation and fertilization choice or through conversion of arable land into permanent grassland,
5. Reduction in crop management emissions through alternative crop mix, fertilizer, irrigation and tillage intensities,
6. Methane reductions through livestock herd size reduction, livestock, manure system improvements, enteric fermentation changes and rice acreage reduction, and
7. Reduction in nitrous oxide emissions from livestock herd size reductions, and alternative fertilization, crop and tillage choice.

Each individual emission and emission reduction category is individually accounted for but also aggregated into a measure of total carbon equivalents. To place different gases on an equal footing, methane and nitrous oxide are converted to carbon equivalents based on the IPCC (1996) 100-year global warming potentials (GWP), which are 21 for methane, 310 for nitrous oxide, and 44/12 for carbon.

ASMGHG can examine the impact of various mitigation policies on the agricultural sector. At each incentive level, it identifies the optimal choice of mitigation strategy. In addition, impacts on the traditional agricultural sector are reported.

3 ECONOMIC FEASIBILITY OF ENERGY CROPS IN THE UNITED STATES

Competitive feasibility of major GHG emission mitigation strategies was simulated by running ASMGHG under a wide range of carbon equivalent prices imposed on net emissions from agriculture. For emissions of non-carbon GHGs, prices were adjusted based on the GWP of the affected GHG relative to carbon. In addition, carbon credits from soil carbon sequestration and afforestation were value-discounted to reflect the saturating nature of these carbon sinks. McCarl and Murray (2001) provide a detailed description along with examination of many alternative set-ups. We chose an average set-up leading to a 25 per cent value discount for sequestered tree carbon and a 50 per cent value discount for sequestered soil carbon. Thus, at a hypothetical carbon price of $20 per TCE, land owners would receive $20 for each ton of offset carbon emissions, $10 for each ton of sequestered soil carbon, and $15 for each ton of carbon sequestered through afforestation, and they would pay for each ton emitted, with $20 for carbon, $114.55 for methane, and $1690.91 for nitrous oxide.

3.1 Major Impacts of Agricultural GHG Emission Mitigation Incentives

Figure 9.1 displays the resultant levels of emission abatement from agricultural mitigation strategies (see also Table 9.1). Unsubsidized energy crops as identified by current technologies are not competitive at zero carbon prices. Economic feasibility of biofuel crops begins at carbon prices around $40 per ton of carbon equivalent. Furthermore, energy crops used as electrical power plant feedstock are more competitive than crops processed into gasoline substitutes such as ethanol.

Under low carbon prices, agricultural management changes, that is tillage changes, are the preferred option. Above carbon incentives of $70 per ton of carbon equivalent, emission offsets from bioelectricity generating crops dominate all agricultural mitigation strategies. For prices above $250 per TCE, the afforestation share increases partially at the expense of bioelectricity generation. For extremely high carbon prices the model favors mitigation options with the highest carbon saving potential per acre, outweighing high operation and implementation costs of these strategies. However, the highest carbon net emission reduction may involve different strategies in different regions.

The results (Table 9.1) also show that large-scale agricultural mitigation efforts divert farmland, reduce traditional crop and livestock production and exports, increase prices, and change agricultural welfare distribution, with producers likely to gain and consumers likely to lose. In addition, GHG mitigation efforts have a mostly positive impact on other environmental externalities related to agriculture such as erosion and water pollution.

3.2 Economic versus Technical Potential

An important concept when regarding biofuel production involves potential to mitigate GHG emissions. Physical scientists often quote very large estimates of potential but these estimates typically neglect the cost of achieving that potential. We used ASMGHG under three settings to derive alternative measures of potential. The first represents technical potential, the second economic potential when considering biofuels as the only mitigation option, and the third competitive economic potential when considering all mitigation options simultaneously.

The technical potential estimate was obtained by changing ASMGHG's objective function from maximizing total economic surplus to maximizing bioelectricity based emission offsets, thereby disregarding economics. This yields substantial emission offsets in the amount of 326 MMTCE annually (Figure 9.2). The single strategy economic potential takes into account all cost of bioelectricity generation except for the opportunity cost related to other possible agricultural mitigation strategies. Under such a setting, a relatively high carbon price of $100 per TCE would only achieve 129 MMTCE emission reduction or 40 per cent of the technical potential. Finally, all agricultural mitigation options were incorporated to find the competitive economic potential. This method yields fewest emission offsets because implementation of other mitigation strategies limits the extent to which biofuel crops can be grown. At $100 per TCE, biofuels would reduce carbon emissions by 115 MMTCE, about 35 per cent of biofuels' technical potential.

The gap in the results between single strategy and competitive economic potential is reflective of inter-strategy competition and relative advantage. Some strategies are superior in some price ranges. The differences between the two economic measures of potential are most notable for high carbon prices (Figure 9.2).

4 SENSITIVITY ANALYSIS ON KEY PARAMETERS OF THE US MODEL

Agriculture in the US is certainly very different from agriculture in Asian Pacific countries. Differences exist with respect to soil, climate, culture-based preferences, per capita land availability, technology and inter-

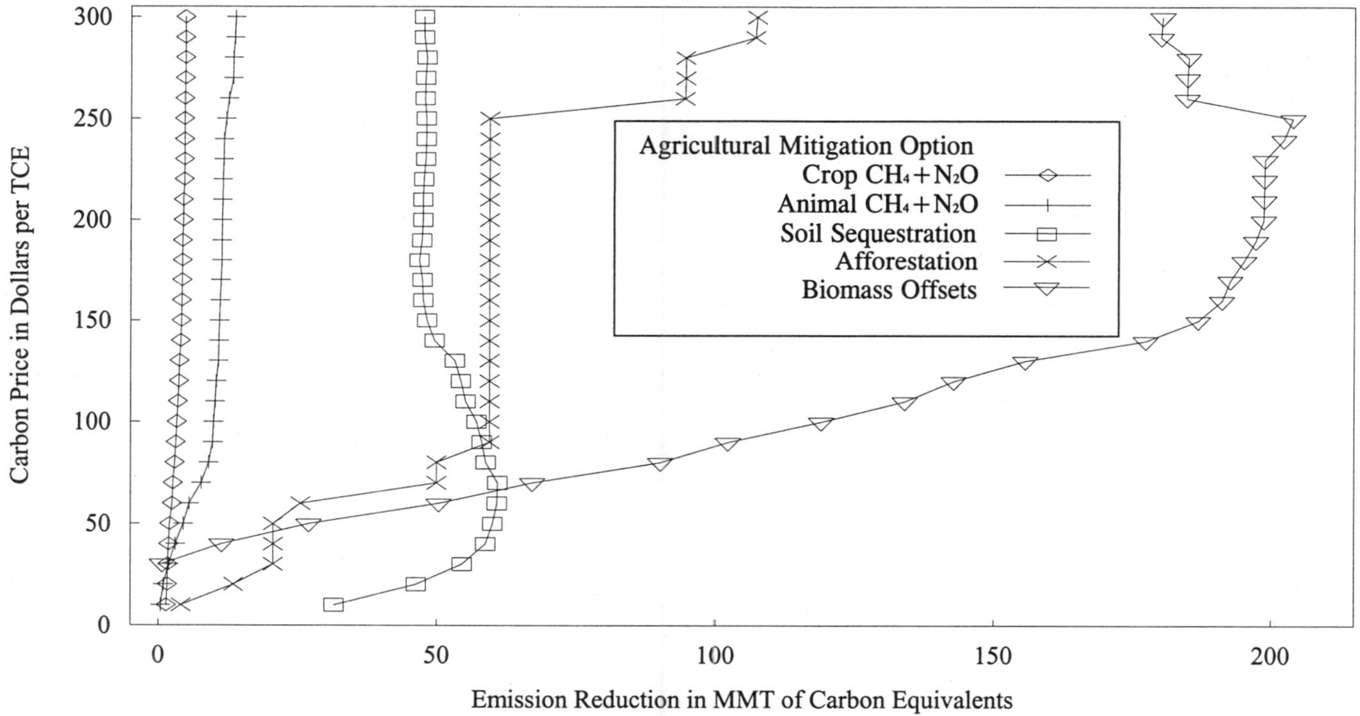

Figure 9.1 Role of major agricultural strategies to greenhouse gas mitigation at selected carbon equivalent prices

Table 9.1 GHG mitigation policy impact summary from the US agricultural sector

Category	Carbon Price in $ per TCE					
	10	20	50	100	200	500
Net GHG Emission Reductions from Major Agricultural Mitigation Accounts (1000 TCE)						
Soil Carbon Sequestration	31540	46316	59945	57105	47520	50291
Afforestation	4028	13445	20619	59407	59407	121278
Electricity from Energy Crops	0	0	27037	118979	198591	172093
Subsidized Corn-Ethanol	2893	2893	2893	2893	2893	2893
Agricultural Inputs	2303	3608	5687	7116	9105	12977
Livestock CH_4 + N_2O	417	875	4428	9826	11301	22330
Crop CH_4 + N_2O	1385	1579	1981	3248	4380	5350
Total Net GHG Emission Reductions from US Agriculture (MMTCE)						
Carbon Dioxide	37.9	63.4	112.2	237.7	306.5	349.7
Methane	0.5	0.8	5.0	13.3	17.9	24.7
Nitrous Oxide	1.4	1.6	2.5	4.7	5.9	9.9
Total Carbon Equivalents	39.7	65.8	119.7	255.7	330.3	384.3
US Farmland Use (Million Acres)						
Traditional Crops	325.06	324.18	313.78	276.6	240.48	229.3
Pasture	395.95	393.22	388.85	373.77	374.24	322.58
Energy Crops	0	0	12.58	57.01	92.66	78.37
Afforestation	0	3.61	5.81	13.63	13.63	37.42
US Agricultural Market Effects (Fisher Index)						
Crop Production	99.43	98.94	96.46	86.23	74.88	69.31
Crop Prices	100.52	101.02	106.36	129.18	162.07	211.84
Crop Net Exports	98.08	96.03	88.60	59.86	31.36	23.69
Livestock Production	99.92	99.62	97.19	91.80	88.47	75.71
Livestock Prices	100.18	100.19	103.77	119.86	142.10	182.87
Agricultural Welfare Changes (Billion $)						
US Farmers' Welfare	0.43	0.6	3.65	13.83	26.59	52.89
US Ag-Consumers' Welfare	−0.39	−0.68	−4.39	−19.46	−41.23	−75.09
Foreign Ag-Sector Welfare	−0.14	−0.25	−1.05	−3.36	−4.81	−5.69
US Ag-Sector Welfare	−0.11	−0.32	−1.8	−8.99	−19.45	−27.89
Other Agricultural Externalities from Traditional Crop Production (% Change per Acre)						
Erosion	−13.1	−25.5	−37.8	−44.3	−49.5	−48.2
Nitrogen Percolation	−8.1	−9.6	−17.0	−18.4	−21.1	−26.3
Nitrogen Subsurface Flow	−10.9	−11.4	−12.7	−11.2	−9.7	−12.3
Phosphorous Loss in Sediment	−23.0	−33.9	−45.2	−50.2	−54.3	−50.1

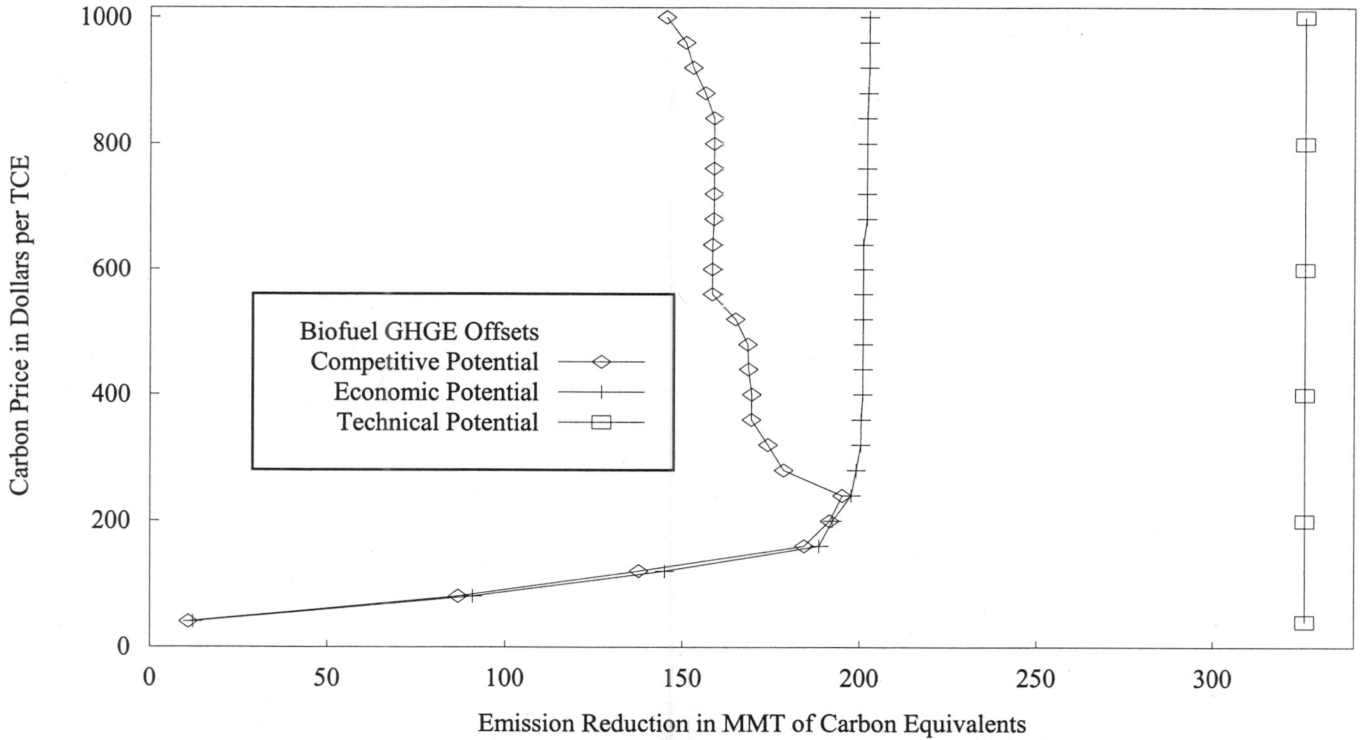

Figure 9.2 Comparison of various measures of carbon emission mitigation potential from energy crops

national market competition (FAO, 2000). To illustrate these differences we conducted a sensitivity analysis on a few decisive model parameters of the US agricultural sector model.

4.1 Biomass Yield

The competitiveness of energy crops depends on many technological parameters. We chose to examine alternative energy crop yields, ranging from 50 to 200 per cent of current US yield estimates. Asian Pacific biomass producers in moist tropical regions may benefit from generally higher plant productivity compared to the US but could also experience lower productivity due to production conditions and altered input mixes. Duke (1983), for example, reports US comparable or higher biomass yields in Asian Pacific countries for eucalyptus and guineagrass (hamilgrass).

Results show a strong impact of energy crop yields on supply of emission offsets (Figure 9.3). In most cases, increased (decreased) yields lead to more than proportional increases (decreases) in biomass emission offsets. For example, at a carbon price of $100 per TCE, energy crops offset about 115 MMTCE per year. A 50 per cent yield increase leads to a 145 per cent increase in emission offsets. In interpreting Figure 9.3 one should keep in mind that all yield increases were implemented without changing crop input parameters. If higher yields were based on higher input use the effects would be less significant.

The results from yield scenarios are not limited to crop yield differences. They can be applied to all technological improvements, from farming to generation of bioenergy which increase the emission input/energy output ratio.

4.2 Energy Price

There are two sources of revenues for producers of energy crops: (a) revenue from selling energy, that is electricity and (b) potential revenue from carbon emission offsets associated with biomass energy. The potential carbon revenue could be a governmental subsidy or an emissions market based income source. If the country is a major energy importer, then its electricity price is determined by the international energy market price. Many Asian Pacific countries, that is, Japan, South Korea, the Philippines, Taiwan and Thailand import substantial amounts of energy (US DOE, 2000). Because currency values differ between countries, the international market price may cause different incentives to grow biomass. For example, agricultural producers in energy-importing countries with a relatively low domestic currency value would earn a higher relative revenue from selling electricity than would US biomass producers. Similarly, revenues would be low where

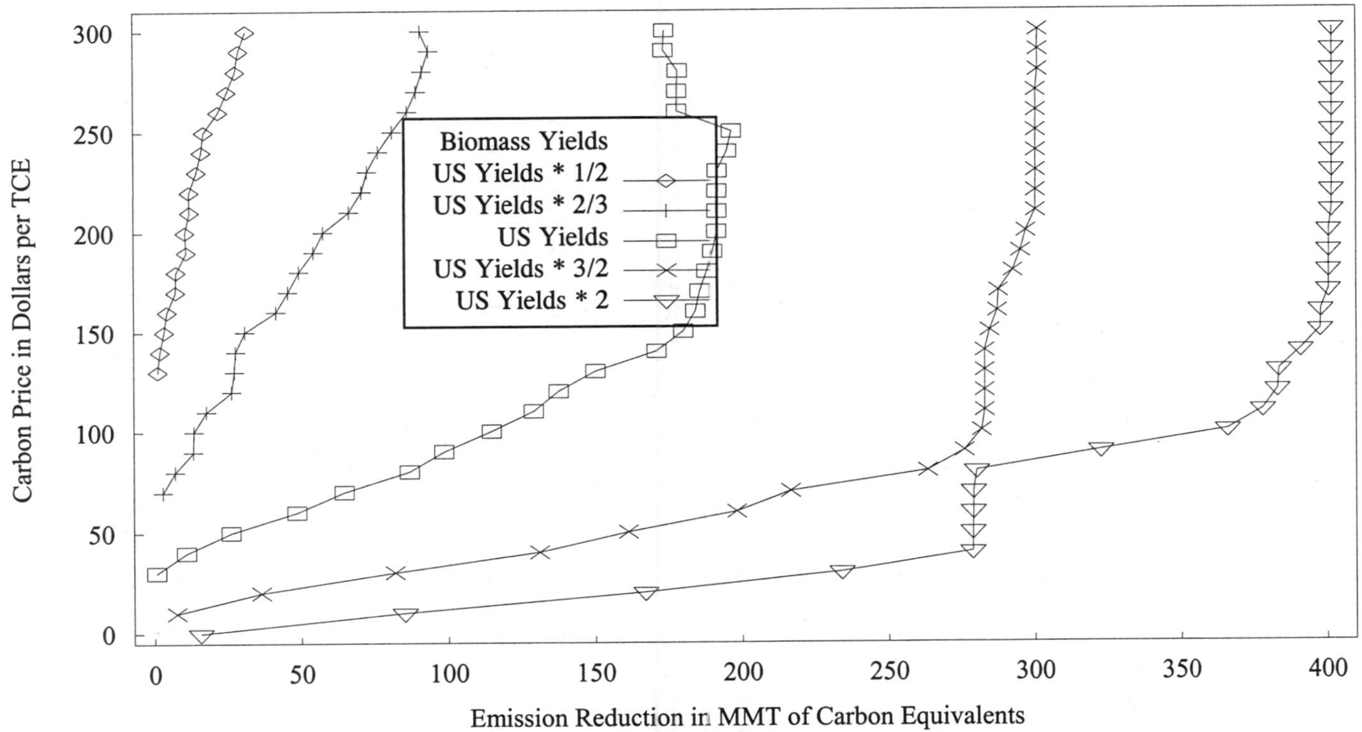

Figure 9.3 *Changes in amount of emission offsets generated by energy crops when energy crop yields are altered*

conventional energy is cheap (energy exporting countries or countries with high currency value). In this study we examined alternative prices for conventional electricity ranging from 50 to 200 per cent of the current US market price.

The effect of energy price changes on the amount of bioelectricity emission offsets is depicted in Figure 9.4. Note that a 100 per cent energy price increase results in much fewer additional emission offsets than does a 100 per cent yield increase. This occurs because selling electricity is only one profit source besides revenues from supplying carbon offsets. If, for example, selling electricity contributes 25 per cent to total profits, then a 20 per cent energy price increase results only in a 5 per cent total profit gain.[1] As the carbon price increases, the electricity revenue becomes relatively less important and energy price changes have less effect on the amount of emission offsets supplied. For example, a doubling of the energy price increases carbon offsets by 54 per cent at $100 per TCE but only by 6 per cent at $150 per TCE (Figure 9.4).

The results on energy price sensitivity are useful beyond extrapolation of US results to foreign countries. They indicate how energy crops' competitiveness changes as fossil fuel prices fluctuate in general. If the recent trend of increased fossil fuel prices continues, energy crop farming may soon be on the rise in the US. Furthermore, the results also indicate how higher or lower costs of energy crop strategies would affect their competitiveness because higher energy prices are equivalent to lower production and processing costs.

4.3 Demand Elasticities

Energy crop production on agricultural land takes away land from traditional agricultural operations. As discussed above, traditional agricultural food and fiber production will lower rising prices. The extent to which prices increase may depend on the elasticity of demand for traditional agricultural commodities. ASMGHG explicitly defines demand curves for 48 primary agricultural products and more than 50 processed products. Each demand curve is specified as constant elasticity function. To assess the effect of higher or lower elasticities, we increased and decreased demand elasticities across all primary agricultural products by a constant fraction ranging from 50 to 200 per cent.

As Figure 9.5 shows, modifications of ASMGHG's demand elasticities have almost no impact on the amount of emission offsets supplied from energy crops. Similar results were obtained when altering export and import elasticities for traded agricultural commodities. Note that elasticities were equally modified across all commodities. In reality, elasticities in foreign countries may be higher or lower depending on the commodity in question.

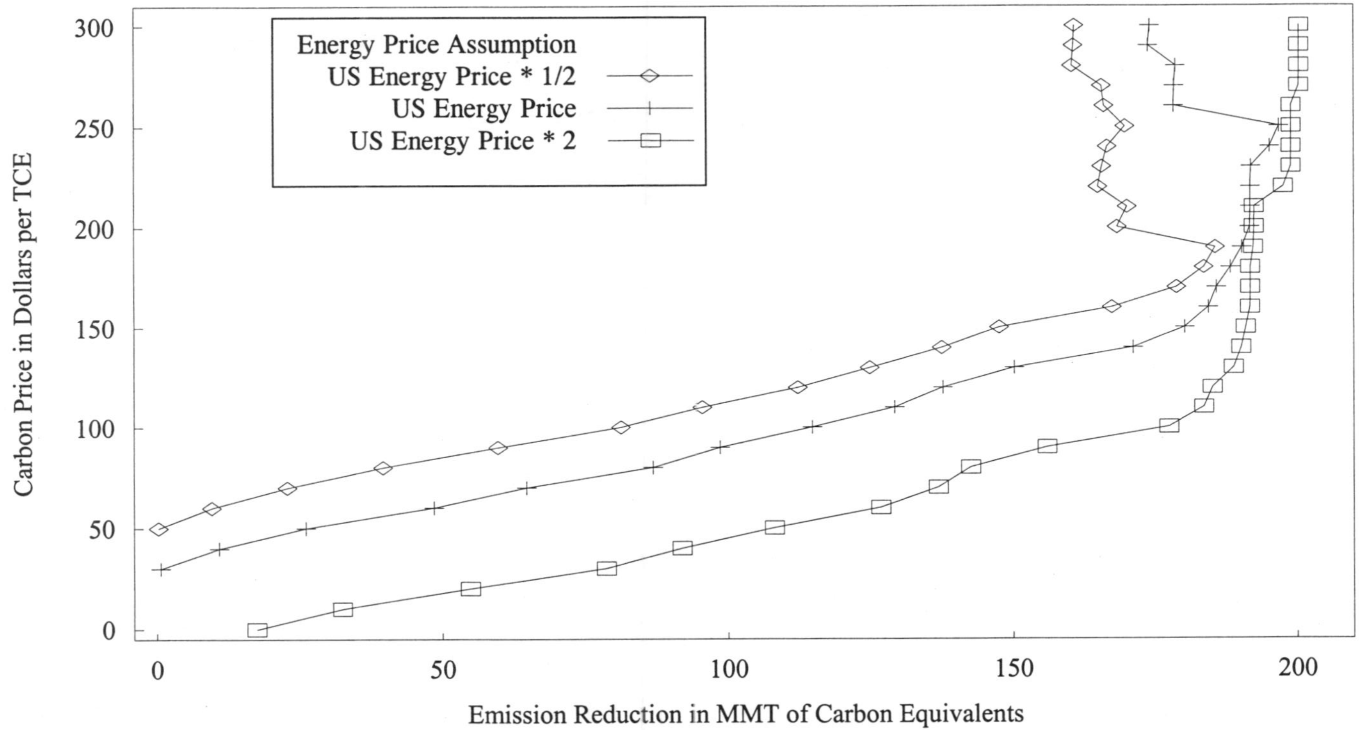

Figure 9.4 Changes in amount of emission offsets generated by energy crops when electricity prices are altered

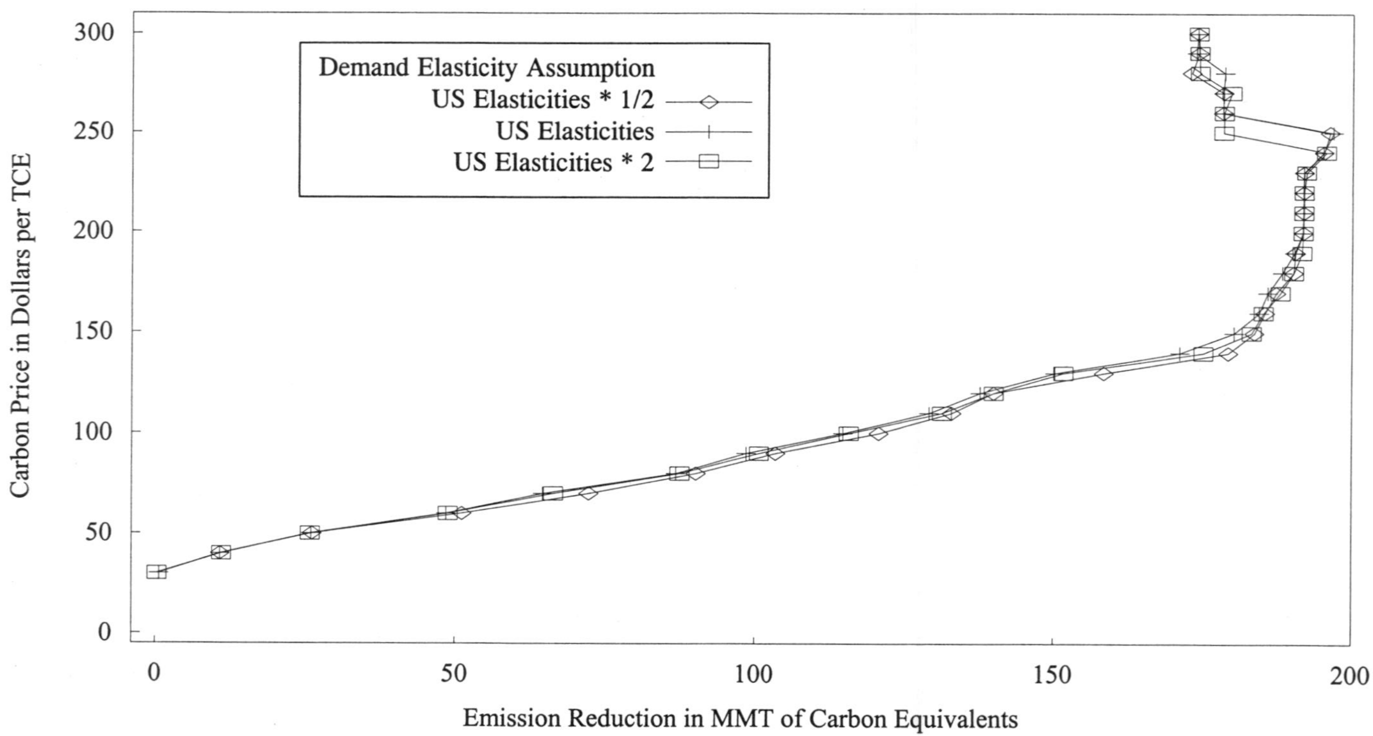

Figure 9.5 *Changes in amount of emission offsets generated by energy crops when domestic agricultural demand elasticities are altered*

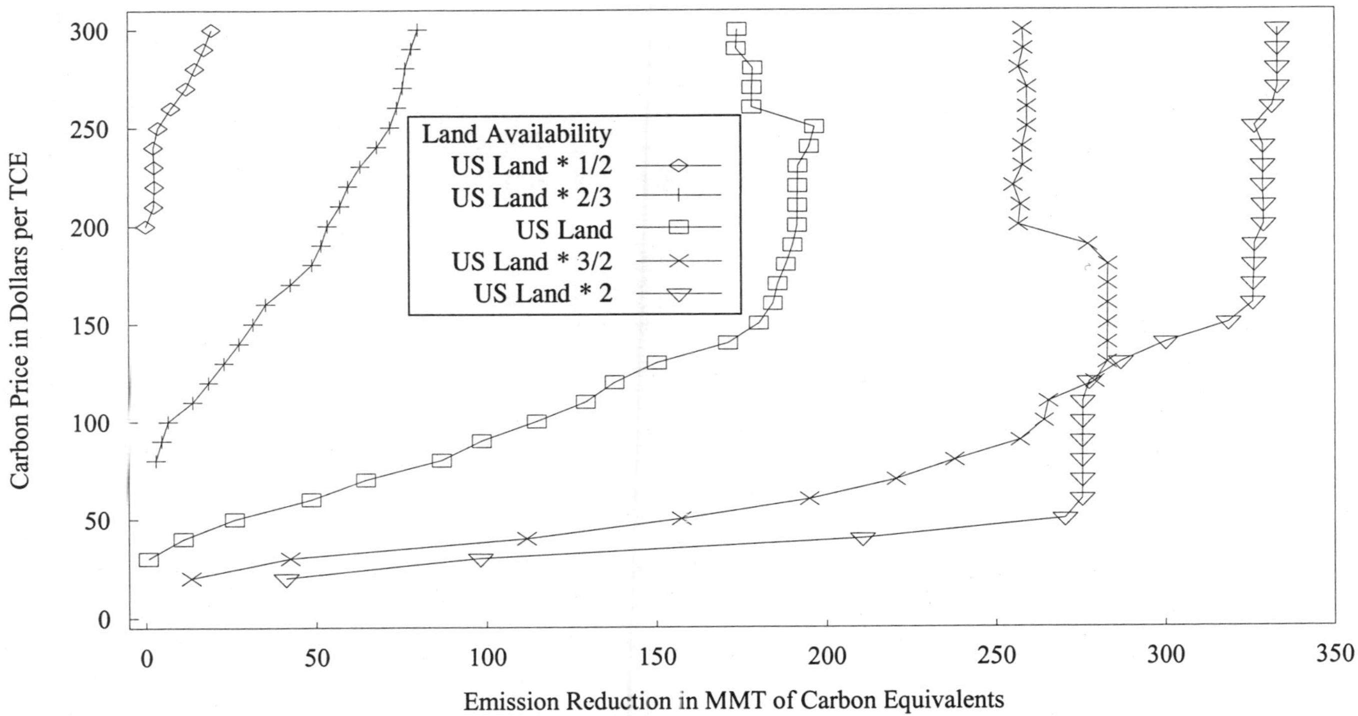

Figure 9.6 Changes in amount of emission offsets generated by energy crops when land availability is altered

4.4 Land Availability

The US has a relatively large agricultural land base relative to its population (FAO, 2000). Taking food and fiber cropland away for the production of energy crops may thus be cheaper in the US but more expensive in densely populated countries, which depend heavily on domestic food production. To illustrate such differences we modified the available agricultural land base in the US between 50 to 200 per cent of its original value.

Figure 9.6 shows supply of bioelectricity emission offsets for different assumptions about the amount of available agricultural land. Not surprisingly, we find energy crops to be very sensitive to land availability. If the US agricultural land base were cut in half, energy crops would not become profitable below carbon prices of \$200 per TCE (Figure 9.6). Less land implies less production, higher commodity prices, and thus higher revenues in the traditional agricultural sector. Consequently, farmers have to give up more by growing energy crops. Limited availability of agricultural land may be a major obstacle to growing energy crops in the Asian Pacific area.

5 CONCLUSIONS

Energy crops supply fossil fuel alternatives and thereby have the potential to reduce greenhouse gas emissions. The attractiveness of energy crops relative to fossil fuels depends on the overall net emission balance and the value of carbon offsets. Economic feasibility implies that the energy output has to be greater than the energy input; otherwise growers could – in the absence of subsidies – not yield a profit. In addition, growing energy crops must be economically superior to other possible GHG mitigation strategies.

Assessments of energy crops in the US agricultural sector show that biomass based electricity (based on switchgrass or short rotation woody crops), while expensive, has considerable potential to offset carbon emissions. Emission offsets range between 1 to 2 metric tons per acre and year. However, a financial support of at least \$40 per ton carbon equivalent (about \$20 per dry ton) is needed to make them economically feasible. Paying less than \$40 per ton of carbon offset induces other agricultural mitigation options, that is changes in tillage or afforestation. Ethanol-generating energy crops turned out to be an inferior strategy over the whole range of analysed carbon prices.

We tried to generalize US results through a sensitivity analysis on key parameters and to extrapolate the findings for Asian Pacific countries. Results indicate that implementation of energy crops in the US is highly sensitive to yields and land availability, sensitive to the price of energy under low and medium carbon prices, but relatively insensitive to demand elasticities of traditional agricultural commodities. With the exception of Australia and Russia, most of the Asian Pacific countries have far less arable land per capita than has the US (FAO, 2000). Allocation of currently cultivated land to energy crops in those countries would imply less land available to produce food. Shortages in domestic food supply, however, could only be offset through increased food imports. Thus, energy crops in most Asian Pacific countries may not be economically feasible unless food imports are cheaper than energy imports.

NOTE

1. Suppose $p=e+c$, where p represents total profit, e represents energy revenue and c represents carbon offset revenue. Suppose further that the ratio of carbon revenue to energy revenue is known, i.e. $c/e=r$. Substitution yields the following identity: $p=e+er=(1+r)e$. If the energy revenue is increased by a factor f, the new profit (np) can be calculated as $np=te+c=te+er=(t+r)e$. Thus, $np/p=(t+r)/(1+r)$. Setting $r=3$, a 20 per cent energy revenue increase ($t=2$) implies a total profit increase of 5 per cent.

REFERENCE

Badin, J. and J. Kirschner (1998), 'Biomass Greens US Power Production', *Renewable Energy World*, **1** (3), pp. 40–45, Nov. 1998.

Chang, C.C., B.A. McCarl, J.W. Mjelde and J.W. Richardson (1992), 'Sectoral Implications of Farm Program Modifications', *American Journal of Agricultural Economics*, **74**, pp. 38–49.

Duke, J.A. (1983), 'Handbook of Energy Crops', unpublished, [Online] available at http:\\www.hort.purdue.edu\newcrop\duke_energy\dukeindex.html.

FAO (Food and Agriculture Organization of the United Nations) (2000), FAOSTAT Statistics Database, 2000, [Online] available at http:\\apps.fao.org.

IPCC (1996), 'Climate Change 1995: The Science of Climate Change. Intergovernmental Panel on Climate Change', in J.T. Houghton, L.G. Meira Filho, B.A. Callander, N. Harris, A. Kattenberg and K. Maskell (eds), Cambridge, UK: Cambridge University Press, pp. 572.

Mann, M.K. and P.L. Spath (1997), *Life Cycle Assessment of a Biomass Gasification Combined-Cycle Power System*, Golden, CO: National Renewable Energy Laboratory, TP-430-23076.

McCarl, B.A. and B.C. Murray (2001), 'Harvesting the Greenhouse: Comparing Biological Sequestration with Emissions Offsets', Department of Agricultural Economics, Texas A&M University, College Station, TX, [Online] available at http:\\ageco.tamu.edu\faculty\mccarl\papers\885.pdf.

McCarl, B.A. and U.A. Schneider (1999), 'Curbing Greenhouse Gases: Agriculture's Role', *Choices*, first quarter, pp. 9–12.

McCarl, B.A. and U.A. Schneider (2000), 'US Agriculture's Role in a Greenhouse Gas Mitigation World: An Economic Perspective', *Review of Agricultural Economics*, **22** (1), pp. 134–59.

McCarl, B.A. and U.A. Schneider (2001), 'Greenhouse Gas Mitigation in US Agriculture and Forestry', *Science*, **294**, 2481–2.

McCarl, B.A., C.C. Chang, J.D. Atwood and W.I. Nayda (2001), 'Documentation of ASM: The US Agricultural Sector Model', Unpublished Report, Texas A&M University. [Online] available at http:\\ageco.tamu.edu\faculty\mccarl\asm.htm

Samson, R. and P. Duxbury (2000), 'Assessment of Pelletized Biofuels', Resource Efficient Agricultural Production – Canada, Quebec, Apr. 2000, [Online] available at http:\\www.reap.ca\Reports\pelletaug2000.html

Schneider, U.A. (2000), 'Agricultural Sector Analysis on Greenhouse Gas Emission Mitigation in the US', PhD Dissertation, Department of Agricultural Economics, Texas A&M University, December.

United Nations (1998), 'Framework Convention on Climate Change', *Kyoto Protocol*, Climate Change Secretariat (UNFCCC), March, [Online] available at http:\\unfccc.int\resource\convkp.html.

US Department of Energy (DOE) (2000), 'Energy Information Administration: Annual Energy Review 1999', September, [Online] available at http:\\www.eia.doe.gov\emeu\aer\

US Department of Energy (DOE) (2001), 'Renewable Energy 2000: Issues and Trends', *Energy Information Administration, Office of Coal, Nuclear, Electric and Alternate Fuels*, February, [Online] available at http:\\www.eia.doe.gov\cneaf\solar.renewables\rea_issues\

Walsh, M.E., D. de la Torre Ugarte, S. Slinsky, R.L. Graham, H. Shapouri and D. Ray (1998), 'Economic Analysis of Energy Crop Production in the US – Location, Quantities, Price and Impacts on the Traditional Agricultural Crops', *Bioenergy 98: Expanding Bioenergy Partnerships*, Madison, Wisconsin, 4–8 October, **2**, pp. 1302–10.

Wang, W., C. Saricks and D. Santini (1999), 'Effects of Fuel Ethanol Use on Fuel-Cycle Energy and Greenhouse Gas Emissions', Center for Transportation Research, Argonne National Laboratory, ANL/ESD-38, January.

Yacobucci, B.D. and J. Womach (2000), 'RL30369: Fuel Ethanol: Background and Public Policy Issues', The National Council for Science and the Environment, Washington, DC, [Online] available at: http:\\www.cnie.org\nle\eng-59.html, March.

PART III

Warming Impacts

10. Climate change and crop yield distribution in Taiwan

Ching-Cheng Chang and Chi-Chung Chen

1 INTRODUCTION

Global warming has been one of the most important issues discussed in recent years. According to the Third Assessment Report of Working Group I of the Intergovernmental Panel on Climate Change (IPCC) (2001), the average global surface temperature has increased by 0.6±0.2℃ over the past century. Precipitation has increased by 0.5 to 1.0 per cent per decade over most mid- and high latitudes of the Northern hemisphere continents, and by 0.2 to 0.3 per cent per decade over tropical land areas. However, in some regions, such as parts of Asia and Africa, the frequency and intensity of droughts have been increasing over the past few decades. The intensity of hurricanes and the increase in the frequency and duration of El Niño events in the last two decades may also be linked to greenhouse-gas-induced global warming (Meejhl and Washington, 1996; Krishnamurti et al., 1998).

Long-term climate variability influences farmers' sowing dates, crop duration, crop yields and management practices, while short-term weather episodes also affect yields by inducing changes in temperature, potential evapotranspiration, and moisture availability. The degree of crops' vulnerability to climate variability depends mainly on the crops' development stage at the time of the weather aberration (Lansigan et al., 2000). Climate variability also influences other factors such as any incidences of pests and epidemic diseases that may hamper crop growth (Luo et al., 1998). Since the magnitudes and the directions of these changes vary across regions and are still uncertain, analyses of the associated effects on world agriculture are highly speculative. Predicted changes in agricultural production must thus be interpreted as being conditional upon the specific changes in the spatial patterns of temperature and rainfall as well as the specific changes over time.

Current climate limitations on crop production and the global pattern of climate changes summarized by the results of crop response models both suggest several different broad geographic zones for which climate change effects on agriculture may differ. Increased precipitation and warming temperatures in high northern latitudes would generally be expected to enhance agricultural production potential in the northern regions of the former Soviet Union, Canada and Europe. Similarly, increased precipitation is predicted to improve agricultural productivity in the southern middle latitudes. On the other hand, drying in the interior of continents in the northern middle latitudes combined with warming temperatures would be expected to lead to negative crop effects in the United States, Western Europe and the most productive regions of Canada (Tobey et al., 1992). As for the low-latitude regions, crop yields are likely to decline because crops are grown nearer their limits of temperature tolerance so that any warming subjects them to higher stress (Rosenzweig et al., 1993). However, many of these results contain uncertainties. They are highly sensitive to the quality and data availability of both the climate prediction and crop simulation models.

Starting from the 1930s, most research on the sensitivity of crop yields to climate changes has concentrated on corn produced in the US. Empirical assessment for other crops produced elsewhere in the world has been quite sparse. Progress has been largely limited, because the collection of field and experiment data is both time-consuming and expensive. With the development of crop growth simulation models during the 1980s and 1990s, it became possible to study the impact of climate change on a wide variety of crops throughout the world. These simulation models have also been applied to study the impact of climate variability and the effect of El Niño on agricultural production (Hoogenboom, 2000). In Asia, a number of modeling studies on the effect of climate change on rice production have emerged since the late 1980s. However, a wide range of predictions has also been made, partly due to the assumptions made in both the climate prediction and the crop simulation models, partly from the use of limited sites for which historical weather data is available, and partly from the complexity in the management practices to cope with these climate effects (Mathews et al., 1997). The controlled experiments, which are the core of the crop simulation models, are expensive and time-consuming to validate and, as argued by Mendelsohn et al. (1996), 'provide poor estimates of the actual magnitude of impacts because they fail to account for the many adjustments that farmers make to environmental conditions'.

There is little evidence available for other tropical and semi-tropical crops grown in the Asian region, as the use of simulation models is still an evolving science in this region. In the analysis that follows, we will use the statistical regression method to examine the distribution of crop yields in response to variations in temperature and precipitation for Taiwan, an economy located on a semi-tropical island. Taiwan grows more than 90 types of vegetables, the most important of which are bamboo shoots, cabbage, vegetable soybeans, tomatoes, Chinese cabbage, watermelon and cantaloupe. In fact, vegetables are second only to rice in terms of acreage sown. Taiwan also has every kind of fruit, from tropical to temperate, available year-round. Bananas have been especially important as an export item, and there have been many improvements in the production and marketing of bananas in recent years. In terms of cultivated area, citrus fruits have been the most important crop. Improvements in cultivation techniques have led to significantly higher quality. Citrus fruits have been popular at home and are also an important export commodity.

This chapter is organized as follows. Section 2 describes the climate characteristics in Taiwan. Section 3 presents the empirical model of crop yield response regressions. In section 4, the implications of alternative climate change scenarios are evaluated and section 5 summarizes the results. The results not only illustrate the sensitivity of crop production to climate change in this area, but also identify the atmospheric conditions that control the yield distributions. Such information can assist us in understanding any possible adaptive behavior that may be available to cope with these changes.

2 CLIMATIC CHARACTERISTICS IN TAIWAN

Taiwan is situated between 21.7° and 25.5° of the northern latitude and has a total area of 3.6 million hectares. A central mountain chain runs along the vertical axis of the island. The landscape rises from a 100-meter foothill to mountain ranges of 3000 meters within a horizontal distance of only 50 to 60 kilometers. Thus, the location of this mountain chain determines the distribution of agricultural land. Total cultivated land in 1996 was 0.87 million hectares, or 24 per cent of the total land area. Of this, paddy fields accounted for 0.46 million hectares and dry land for 0.41 million hectares. About 75 per cent of all the paddy fields were irrigated.

The sub-tropical weather in Taiwan permits the growing of a great variety of crops. In the summer months, the average temperature in southern and northern Taiwan is almost identical (27–28℃), but during the winter months, the mean temperature is about 5℃ higher in the south (20℃) than in the north (15℃). Taiwan's prevailing wind during the winter and most of the spring is called the Northeast Monsoon. During the transition period from spring to summer, the Mei-yu[1] climate prevails as the Northeast Monsoon gives way to the southeasterly wind. Because Taiwan is surrounded by sea, the Northeast Monsoon and the Mei-yu are the two important weather systems in Taiwan.

Other tropical climatic phenomena, such as the El Niño Southern Oscillation (ENSO) and typhoons, are also critical to agricultural production in the region. Although Taiwan is located outside ENSO's maximum effective zone, there are strong indications that ENSO is related to East Asia's monsoon strength (Zhang et al., 1996). Hsu et al. (2001) also report that ENSO is closely related to Taiwan's winter climate, in particular, the variations in winter temperature. Typhoons, characterized by heavy rainfall and strong winds, are also destructive to many crops. Therefore, like all her Asian neighbors, Taiwan's crop production is highly susceptible to climate variability. Floods and droughts resulted in US$1.8 billion in crop losses[2] during 1990–99, which was about 4 per cent of the total crop values produced during the 10-year period.

Although rainfall distribution is profoundly affected by topography, the annual rainfall in the main crop production area ranges from 1500 mm to 2000 mm. Rainfall is quite abundant throughout the island from May to October. During the winter months, the rainy season prevails in the northern and eastern regions while the central and southern parts of the west coast have dry and sunny weather. Thus, given the warm and humid weather and adequate irrigation water, many kinds of crops can be grown to produce a wide variety of farm products on the island (Cheng, 1975).

As shown in Figures 1 and 2 in Chang (2001), regional differences in annual average temperature and precipitation are presented. Over the past 20 years, the temperature variation has obviously been less dramatic than the variation in precipitation. Both the southern and eastern regions have above average temperature, while their precipitation levels are mostly below average.

3 YIELD REGRESSION MODELS

Changes in weather not only have an effect on the mean and variance of crop yields (Thompson, 1986), but they can also affect the higher moment of crop yield distributions. For example, Gallagher (1986) and Kaylen and Koroma (1991) find that the distribution of US corn yields was not only related to climate, but also negatively skewed. In our study, higher moments other than the sample mean (μ) and variance (σ^2) are used to characterize the yield distribution and are regressed using a set of climate variables. The measure of skewness is estimated as the ratio of the third moment to the cube of the standard deviation (i.e. μ_3/σ^3, where μ_i denotes the ith moment). The measure of kurtosis is calculated as the ratio of the fourth moment to the variance squared (i.e. μ_4/σ^4). Because the kurtosis of the normal distribution has an expected value of 3, a measure of excess kurtosis, $(b_2 - 3)$, is often reported in empirical work (Judge et al., 1982, p. 891).

Temperature and precipitation are considered to be the major climate variables. Crop growth is connected to the climate factors via a combination of linear and non-linear responses. Non-linearity of response means that it is necessary to preserve the variability of weather sequences in empirical estimations (Semenov and Porter, 1995). In addition, weather variation should be included to reflect the influence of departures from normal weather conditions on crop yields. For example, Thompson (1986) reported that decreasing weather variability was favorable for US corn yields before 1970. Using a mechanical crop model and farm-level yields, Park and Sinclair (1993) found that variations in temperature and precipitation affected the distribution of US corn yields over time and space. Mendelsohn et al. (1999) showed that climate variations had significant impacts on farming in the US Increases in interannual climate variation were predicted to be harmful to US agriculture, but decreases in diurnal variation are beneficial. Overall, omitting the variation terms tends to underscore the beneficial effect from global warming.

To reflect the response of yield distribution, $F(Y)$, to the two climate variables (temperature (T), precipitation (R)) as well as their variations, the regression model is given by:

$$F(Y) = f(T, R; T^2, R^2, T * R; VT, VR)$$

where T^2, R^2, and $T * R$ denote the non-linear components of temperature and precipitation, while VT and VR denote the variations in monthly mean temperature and precipitation from their 20-year monthly averages, respectively. They are included in order to evaluate the changes in weather variability on yield.

Because the model adopts a non-linear specification for each climate variable where linear and quadratic terms are used as regressors, it reflects the effect of a physiological optimum on yield. For example, a positive coefficient for the linear term and a negative coefficient for the quadratic term indicate that an intermediate value has the greatest positive effect on yield distribution (Mendelsohn et al., 1994, 1996; Kaufmann and Snell, 1997). This also allows for a non-monotonic relationship between climate and the shape of yield distribution, for example, a warming up might increase crop yields in cooler areas, but decrease yields in warmer regions (Segerson and Dixon, 1999). However, co-linearity problems arise from the inclusion of these quadratic terms. Therefore, the results should be interpreted carefully as the significance of the coefficient estimates might be unreliable.

The distributions of rice, sweet potato, tomato, ponkan mandarin (Citrus reticulata Blanco) and banana yields are illustrated by a pooled time-series and cross-sectional data of 15 regions over the period 1977–96. Both the spatial and temporal distributions of yields are calculated. Therefore, the effects from weather and weather variability can be measured across years and across locations. For spatial distributions, a time-trend variable (*TIME*) is also added to represent the effect of technological progress during the sample period, which can be attributed to increasing fertilizer application, new high-yielding crop varieties, improved cropping practices, and so on. The main purpose of adding this time-trend variable is to capture the contribution from technology and management improvements.

4 REGRESSION RESULTS

A semi-logarithmic functional form is used where all the independent variables (except the time-trend variable) are transformed by taking their logs. The data on crop yields are drawn from the *Taiwan Agricultural Yearbook*. Monthly weather data on temperature and precipitation are obtained from Taiwan's Central Weather Bureau. The ordinary least squares method is used in both over-time and over-space regressions. The summary statistics are presented in Table 10.1, and indicate that the spatial distributions of the five selected crop yields tend to be more skewed than their temporal distributions. The lower kurtosis in relation to rice and banana yields indicates more flat-topped yield distributions than the other three crops. As for the climate variables, the variations in precipitation are more dramatic than the variations in temperature in both space and time.

The regression results for each crop are listed in Tables 10.2–10.6, respectively.

4.1 Rice

Rice is Taiwan's most important staple crop. Table 10.2 reports that the model shows pretty good explanatory power with regard to the mean yield as measured by the adjusted R^2. However, the explanatory powers of the other moments are very poor. The results also indicate that precipitation has no effect on any of the yield distribution components. Since more than 80 per cent of Taiwan's rice paddies are irrigated and the irrigation systems are reportedly among the most efficient ones in the world (Lam, 1996), precipitation is not as critical in relation to yield distributions as the other climate factors.

For each region, the mean yield over time (Table 10.2) is significantly affected by temperature variations as well as the offsetting effect from precipitation variations. Temperature variations are also found to have a significant influence on the mean yield over space (Table 10.2). These findings are in accordance with those reported by Lansigan et al. (2000), which indicate that rice yields in the Philippines are directly influenced by both short-term weather fluctuations and longer-term climate variability. In addition, technology, as expected, has a positive and significant effect on mean yield over space.

Table 10.1 Summary statistics of regression data

Variable	Over Time (observations=15)				Over Space (observations=20)			
	Mean	Std Dev	Minimum	Maximum	Mean	Std Dev	Minimum	Maximum
I. Dependent Variable:								
1. Rice yield (kg/hectare)								
Mean	4509.98	512.95	3480.75	5219.80	4509.98	814.69	3170.53	5643.67
Std Dev	858.53	70.59	734.37	970.29	572.70	123.65	363.48	801.11
Skewness	−0.24	0.41	−1.05	0.65	−0.39	0.32	−0.91	0.15
Kurtosis	−0.89	0.39	−1.44	−0.04	−0.20	0.77	−1.24	1.43
2. Sweet potato yield (kg/hectare)								
Mean	15747.52	4453.64	10055.40	27483.20	15747.52	1138.88	14156.53	18708.27
Std Dev	2085.43	1241.50	664.69	5396.85	4855.77	933.08	3005.92	6206.34
Skewness	0.55	1.18	−0.73	3.79	1.08	0.44	0.19	1.79
Kurtosis	1.19	4.45	−1.81	15.73	1.35	1.39	−1.04	4.01
3. Banana yield (kg/hectare)								
Mean	18269.67	6725.01	10826.80	30325.90	18269.67	1703.36	15224.20	21550.40
Std Dev	4106.31	2725.48	953.95	9539.15	8093.00	1023.70	5651.96	10388.62
Skewness	−0.08	0.75	−0.95	1.40	0.77	0.42	−0.02	1.33
Kurtosis	0.04	1.31	−1.58	2.40	−0.57	0.73	−1.52	0.86
4. Ponkan yield (kg/hectare)								
Mean	11465.11	3170.99	6914.80	19244.55	11465.11	901.60	9802.73	12829.93
Std Dev	2292.39	844.16	1285.98	4781.41	3857.12	629.35	2862.62	5342.06
Skewness	−0.18	1.12	−2.25	1.82	0.85	0.55	0.09	2.23
Kurtosis	1.59	2.61	−1.50	6.72	1.29	1.95	−1.15	7.20
5. Tomato yield (kg/hectare)								
Mean	20502.04	9363.75	11079.50	47616.60	20502.04	2563.30	15657.67	23400.27
Std Dev	5245.29	2981.36	1380.31	11853.73	10375.74	3209.30	5665.31	17594.63
Skewness	0.29	1.24	−2.07	2.37	1.48	0.45	0.58	2.18
Kurtosis	1.72	3.06	−1.47	8.40	3.10	1.80	0.80	6.93
II. Explanatory Variables:								
T (℃):	22.80	1.90	16.72	24.96	22.80	0.25	22.33	23.25
R (mm.)	189.35	82.04	134.51	401.07	189.35	26.41	124.19	223.09
VT	20.96	5.69	8.60	27.81	20.96	2.92	16.55	27.42
VR	41627.17	19172.77	22822.63	83565.61	41627.17	16661.83	12568.63	78537.99

In the over-space yield distribution regression results in Table 10.2, an increase in temperature tends to reduce the mean and variance, but it is not statistically significant. Nevertheless, a higher temperature tends to increase the skewness of the yield distribution across different regions in Taiwan, while decreasing the kurtosis in a significant manner. This suggests that a warmer climate leads to a flatter, but less symmetric, yield distribution function for rice. Therefore, greater uncertainty with regard to the rice yield is expected as the climate becomes warmer.

In a cross-country comparison of five Asian countries, Luo et al. (1998) investigate the risks of yield loss caused by rice leaf blast associated with global warming. They report that a temperature change has different effects on yield loss in different countries. In general, the risk of yield loss caused by blast epidemics in the humid sub-tropical areas is higher than in the sub-humid sub-tropics. However, the results are very sensitive to the cultivation systems and the quantity of nitrogen and water supply. Matthews et al. (1997) also conduct a crop simulation study for rice produced in the Asian region. Their results show that China, Thailand, Bangladesh and Japan will experience a decline in production while Indonesia, Malaysia, Taiwan and parts of India and China are predicted to benefit from global warming. After taking into account the adaptation in management practices and new varieties, the potential for increased rice production in the region is expected. In comparison, our results suggest that although warming may not lead to yield losses for rice produced in warm, humid regions, it may affect the higher moments of the yield distributions, indicating a change in the probability of yields being found near the upper or lower limits.

4.2 Sweet Potato

The sweet potato used to be a very important staple crop of rural households as well as a feed ingredient for hogs in Taiwan. Its production value accounted for more than 10 per cent of total crops produced in 1960. Its importance has fallen dramatically and now its production value is less than 1 per cent of all crops. Nevertheless, the sweet potato's nutrition and fiber contents are highly regarded by many health-conscious city dwellers.

As shown in Table 10.3, the overall explanatory power of the first moment of the sweet potato's yield distribution is quite high. Some of the other three moments also have very high R-squares. Therefore, climate variables are influential factors in explaining its yield performance.

For the over-time yield distribution (Table 10.3), climate and climate variability are found to have a significant and non-monotonic impact on the temporal distribution of sweet potato yields. In other words, increases in temperature and precipitation can be beneficial over some ranges, but detrimental over the others. Based on the results in the second moment, increases in temperature and precipitation also lead to higher temporal variations in sweet potato yields.

As for spatial distribution (Table 10.3), a rise in temperature has a positive and significant impact on mean yield, while variations in temperature

Table 10.2 Relationship between Taiwan's climate variables and the distribution of rice yields: 15 counties, 1977–96

	Over time					Over space			
	Mean	SD	SK	KU		Mean	SD	SK	KU
Constant	13185	−14164	−55.48	−112.47	Constant	312632	35945	−1122.14	2754.55
	(74801.57)	(24900.17)	(123.04)	(131.66)		(380660.38)	(270762.93)	551.83	1420.89
					Year	123.71***	−2.67	0.01	−0.01
						(9.76)	(6.94)	(0.01)	(0.04)
T	1094.44	1056.03	2.96	6.91	*T*	−52599	−4996.18	92.59*	−241.49*
	(5559.67)	(1850.72)	(9.15)	(9.78)		(32178.41)	(22888.44)	(46.64)	(120.11)
T^2	−50.49	−19.25	−0.036	−0.10	T^2	1231.27*	157.62	−1.95*	5.29*
	(102.79)	(34.22)	(0.017)	(0.18)		(687.58)	(489.07)	(0.99)	(2.57)
R	−100.56	30.53	0.16	0.24	*R*	387.21	236.77	0.48	−0.07
	(106.47)	(35.44)	(0.18)	(0.18)		(291.77)	(207.53)	(0.42)	(1.08)
R^2	0.075	−0.02	−0.00	−0.00	R^2	−0.09	−0.03	−0.00	−0.00
	(0.04)	(0.016)	(0.00)	(0.00)		(0.07)	(0.05)	(0.00)	(0.00)
*T*R*	2.14	−1.01	−0.01	−0.01	*T*R*	−15.22	−9.72	−0.02	0.01
	(4.11)	(1.38)	(0.01)	(0.01)		(12.24)	(8.07)	(0.02)	(0.05)
VT	−11135**	8.33	0.06	0.09	*VT*	71.41***	15.06	0.03	−0.04
	(41.57)	(13.84)	(0.07)	(0.07)		(17.78)	(12.65)	(0.03)	(0.07)
VR	0.048*	0.005	−0.00	−0.00	*VR*	−0.007	−0.005	−0.00	0.00
	(0.021)	(0.007)	(0.00)	(0.00)		(0.005)	(0.003)	(0.00)	(0.00)
Deg. of Freedom	14	14	14	14	Deg. of Freedom	19	19	19	19
R-squared	0.8818	0.3084	0.4997	0.3587	R-squared	0.9709	0.3611	0.6018	0.5441
Adj R-squared	0.7636	−0.3832	−0.0006	−0.2826	Adj R-squared	0.9498	−0.1035	0.3121	0.2125

Notes: Numbers in parentheses are standard errors.
* significant at 1% level.
** significant at 5% level.
*** significant at 10% level.

offset some of this positive effect. Technological progress is found to be favorable to the variance only, while no association is found with the other moments of the spatial distribution.

4.3 Banana

During the 1960s–70s, bananas were one of the most important commodities exported to Japan from Taiwan. They are also a very important cash crop for farmers in the central and southern regions. Although the export market has been taken over by the Philippines, the popularity of bananas in the domestic market remains stable. As shown in Table 10.4, neither temperature nor rainfall has any significant impact on the mean yield. However, they both do have a significant impact on the second moment of the over-time and the over-space distributions, but in an opposite direction. This indicates that warming increases the temporal yield variations while reducing spatial variations. The two climate variables also have some impacts on the skewness and kurtosis. This highlights the importance of looking into the higher moments of yield distributions while studying the relationship between climate and yield.

Another interesting result for bananas is that technological progress not only enhances the mean as expected, but also makes the shape of the yield distribution symmetrical. Variability in rainfall also contributes significantly to the second moments, which indicate that the yield distribution will widen as rainfall becomes more unpredictable.

4.4 Ponkan Mandarin (Citrus reticulata Blanco)

The climate impacts on citrus fruit vary by species and regions. The yield changes used in Adams et al. (1999) (originally estimated by Ben Mechlia and Carroll (1989a,b) using the crop simulation models) for citrus (Valencia orange) produced in the US show that warming temperatures may cause some decreasing yields in the southern regions, while showing increasing yields for more northern sites. Rosenzweig et al. (1996) find a similar negative impact from warming for the most southern regions of the US, adding that higher CO_2 levels may counteract the negative impact. However, they also conclude that more work is needed to improve the existing models, because their results are poorly correlated with the observed yields. Using Taiwan's historical data and statistical methods, Chang (2001) reports that a temperature rise is favorable for ponkan, tankan and wentan, but not for liucheng, lemons and grapefruit planted in Taiwan.

Ponkan mandarin is the most important citrus crop in Taiwan, followed by liucheng. From Table 10.5, the simple regression method cannot produce satisfactory results for all cases. The adjusted R-squared values are in gen-

Table 10.3 Relationship between Taiwan's climate variables and the distribution of sweet potato yields: 15 counties, 1977–96

	Over Time					Over Space			
	Mean	SD	SK	KU		Mean	SD	SK	KU
Constant	−2256545	−730205	317.24	813.15	Constant	−1406608	35308	1272.09	3339.69
	(669784.24)	(330628.45)	(234.301)	(637.37)		(895708.53)	(1279836.36)	(948.25)	(2628.94)
					Year	169.13	123.49***	−0.01	−0.03
						(22.96)	(32.81)	(0.02)	(0.07)
T	173936***	53865*	−21.49	−58.96	*T*	95698***	−23497	−109.13	−286.99
	(49782.07)	(24574.14)	(17.41)	(47.37)		(75717.04)	(108188.57)	(80.16)	(222.23)
T^2	−3326.84***	−959.86*	0.36	1.08	T^2	−2101.15	509.09	2.40	6.29
	(920.48)	(454.38)	(0.32)	(0.88)		(1617.89)	(2311.74)	(1.71)	(4.75)
R	2963.95**	918.41*	−0.55	−1.12	*R*	11.56	−56.49	0.06	0.09
	(953.38)	(470.62)	(0.33)	(0.91)		(686.54)	(980.96)	(0.73)	(2.01)
R^2	−0.88*	0.05	0.00	0.00	R^2	−0.018	0.06	0.00	0.00
	(0.43)	(0.21)	(0.00)	(0.00)		(0.15)	(0.22)	(0.00)	(0.00)
*T*R*	−120.43**	−39.91*	0.02	0.04	*T*R*	−0.71	1.11	−0.00	−0.01
	(36.79)	(18.16)	(0.01)	(0.04)		(28.79)	(41.13)	(0.03)	(0.08)
VT	527.75	254.47	−0.27*	−0.45	*VT*	−90.47*	−71.89	0.04	0.21*
	(372.23)	(13.74)	(0.13)	(0.35)		(41.86)	(59.81)	(0.04)	(0.12)
VR	0.46**	−0.16	−0.00	0.00	*VR*	0.01	0.01	0.00	−0.00
	(0.19)	(0.09)	(0.00)	(0.00)		(0.01)	(0.02)	(0.00)	(0.00)
Deg. of Freedom	14	14	14	14	Deg. of Freedom	19	19	19	19
R-squared	0.8743	0.6058	0.7802	0.8866	R-squared	0.9176	0.7493	0.3826	0.5232
Adj R-squared	0.7486	0.2116	0.5603	0.7724	Adj R-squared	0.8576	0.567	−0.0665	0.1765

Notes: See Table 10.2.

eral lower than those for the other crops. Furthermore, we cannot identify any significant climate factor. However, technological progress is the only factor that is beneficial to mean yield.

Similar to other citrus plants, ponkan is very sensitive to extreme temperatures, which are not included in our explanatory variables. In addition, because ponkan is perennial, climate variability not only affects its current yields, but also its future states. A dynamic relationship may be a crucial feature in the model specification. Like the conclusion in Rosenzweig et al. (1996), more work needs to be done to obtain a better model for citrus.

4.5 Tomato

Tomatoes are consumed as a vegetable as well as a fruit in the diet of Taiwanese. They are also processed into juice and ketchup for domestic consumption and export. We choose the tomato because the fruit is sensitive to heat during the ripening stage. Because greenhouse production is very popular in many countries, the relationships between climate and tomato yields are not widely studied.

In Chang (2001), Taiwan's tomato yields are found to be positively related to a temperature increase. This is quite similar to the pattern used in the Adams et al. (1999) study for the US. From the R-squared value in the over-time regressions in Table 10.6, climate-related variables are indeed very important factors in determining the first and second moments of the distribution over time. In particular, both the mean and variance are responsive to a temperature change in a non-linear fashion. As for the distribution over space, none of the climate factors matter, but technological change is the most important contributor to yield growth and its geographic variations.

Rainfall is also found to be an influential factor in a similar non-linear fashion in the over-time regression. Similar to the case of bananas, variability in rainfall widens the distribution of yields, which suggests that low yields will become more likely as rainfall becomes more variable.

5 CONCLUDING REMARKS

This chapter addresses the issues involving the extent of crop vulnerability to global warming and the associated climate variability. We investigate the impact of climate change on the yield distributions of five semi-tropical crops in Taiwan. Two climate variables (temperature and precipitation) along with their intra-annual variations are considered in our regression models.

Our empirical results show that the two climate variables have a significant and non-monotonic impact on the mean yields as well as other moments

Table 10.4 Relationship between Taiwan's climate variables and the distribution of banana yields: 15 counties, 1977–96

	Over time					Over space			
	Mean	SD	SK	KU		Mean	SD	SK	KU
Constant	1431291	−1048789	−109.72	525.48	Constant	−2180570	2941335	1122.71	1130.61
	(1783082.78)	(566301.17)	(183.42)	(327.16)		(2830897.32)	(1457575.54)	(585.76)	(1329.15)
					Year	223.19**	−5.09	−0.06***	−0.05
						(72.58)	(37.37)	(0.02)	(0.03)
T	−103362	80197*	5.98	−44.37	*T*	147927	−225770*	−81.19	−77.39
	(132528.59)	(42090.64)	(13.63)	(24.32)		(239304.59)	(123213.41)	(49.51)	(112.36)
T^2	1873.91	−1512.04*	−0.08	0.89*	T^2	−3096.99	4312.61	1.62	1.43
	(2450.47)	(778.26)	(0.25)	(0.45)		(5113.38)	(2632.78)	(1.06)	(2.40)
R	−1731.52	1436.95*	0.37	−0.28	*R*	582.14	−3663.36***	−0.88*	−1.48
	(2538.07)	(806.08)	(0.26)	(0.47)		(2169.81)	(1117.19)	(0.45)	(1.02)
R^2	0.03	−0.33	−0.0003**	−0.00	R^2	0.25	0.56**	0.00	0.00
	(1.14)	(0.36)	(0.00)	(0.00)		(0.48)	(0.25)	(0.00)	(0.00)
*T*R*	72.69	−53.84*	−0.01	0.02	*T*R*	−29.47	152.25***	0.04*	0.06
	(97.95)	(31.11)	(0.01)	(0.02)		(90.99)	(46.85)	(0.02)	(0.04)
VT	−1180.39	−169.24	0.13	0.10	*VT*	24.94	24.96	−0.02	−0.08
	(990.93)	(314.72)	(0.10)	(0.18)		(132.29)	(68.11)	(0.03)	(0.06)
VR	0.17	−0.29*	−0.00	0.00	*VR*	0.01	−0.05***	−0.00	0.00
	(0.51)	(0.16)	(0.00)	(0.00)		(0.03)	(0.02)	(0.00)	(0.00)
Deg. of Freedom	14	14	14	14	Deg. of Freedom	19	19	19	19
R-squared	0.6655	0.6533	0.6093	0.76	R-squared	0.632	0.7299	0.743	0.5531
Adj R-squared	0.331	0.3065	0.2185	0.5201	Adj R-squared	0.3643	0.5334	0.5661	0.2281

Notes: See Table 10.2.

Table 10.5 Relationship between Taiwan's climate variables and the distribution of ponkan yields: 15 counties, 1977–96

	Over time					Over space			
	Mean	SD	SK	KU		Mean	SD	SK	KU
Constant	209507	−398092	−22.58	1002.61	Constant	−1940834	5830.15	922.64	2331.85
	801584.28	281466.68	282.11	847.09		(1631022.38)	(1621134.75)	(1099.56)	(3781.65)
					Year	77.28*	−26.52	0.02	−0.01
						(41.82)	(41.56)	(0.03)	(0.09)
T	−14001	30205	−0.05	−74.16	*T*	158237	−3229.11	−87.82	−225.12
	(59578.18)	(20920.16)	(20.97)	(62.96)		(137875.42)	(137039.59)	(92.95)	(319.67)
T^2	259.97	−551.49	0.02	1.93	T^2	−3461.06	231.0	1.97	5.38
	(1101.61)	(386.82)	(0.39)	(1.16)		(2946.08)	(2928.22)	(1.99)	(6.83)
R	−299.33	455.65	0.16	−1.56	*R*	−7.47	791.35	0.27	2.55
	(1140.99)	(400.64)	(0.41)	(1.21)		(1250.14)	(1242.56)	(0.84)	(2.89)
R^2	0.22	0.07	−0.00	0.00	R^2	0.15	−0.06	−0.00	−0.00
	(0.51)	(0.18)	(0.00)	(0.00)		(0.28)	(0.27)	(0.00)	(0.00)
*T*R*	8.11	−20.47	−0.01	0.06	*T*R*	−2.86	−33.17	−0.01	−0.10
	(44.03)	(15.46)	(0.02)	(0.05)		(52.42)	(52.11)	(0.03)	(0.12)
VT	114.33	54.62	0.21	−0.33	*VT*	−106.69	−2.04	0.09*	0.26
	(445.47)	(156.42)	(0.16)	(0.47)		(76.22)	(75.76)	(0.05)	(0.18)
VR	−0.09	−0.09	0.00	−0.00	*VR*	0.02	−0.01	−0.00	−0.00
	(0.23)	(0.08)	(0.00)	(0.00)		(0.02)	(0.02)	(0.00)	(0.00)
Deg. of Freedom	14	14	14	14	Deg. of Freedom	19	19	19	19
R-squared	0.6448	0.3821	0.6491	0.4165	R-squared	0.5639	0.1159	0.4674	0.4975
Adj R-squared	0.2896	−0.2359	0.2982	−0.167	Adj R-squared	0.2468	−0.5271	0.08	0.132

Notes: See Table 10.2.

Table 10.6 *Relationship between Taiwan's climate variables and the distribution of tomato yields; 15 counties, 1977–96*

	Over Time					Over Space			
	Mean	SD	SK	KU		Mean	SD	SK	KU
Constant	−3975497	−694324	−136.84	652.26	Constant	3696098	3696098	1316.46	1939.65
	(1051704.33)	(434204.67)	(317.22)	(1064.90)		(3427348.70)	(3427348.70)	(1006.28)	(3771.55)
					Year	255.27**	255.27**	0.01	0.19*
						(87.87)	(87.87)	(0.03)	(0.09)
T	306914***	63411*	10.92	−44.37	*T*	−388127	−388127	−113.75	−188.91
	(78168.49)	(32272.49)	(23.58)	(79.15)		(289724.49)	(289724.49)	(85.06)	(318.82)
T^2	−5948.03***	−1415.73**	−0.21	0.75	T^2	8929.15	8929.15	2.42	3.88
	(1445.35)	(596.72)	(0.44)	(1.46)		(6190.74)	(6190.71)	(1.82)	(6.81)
R	5699.00***	475.53	0.14	−1.36	*R*	2293.81	2293.81	−0.34	−1.14
	(1497.01)	(618.05)	(0.45)	(1.52)		(2626.97)	(2626.97)	(0.77)	(2.89)
R^2	−2.59***	−0.20	0.00	0.00	R^2	−0.50	−0.5	0.000	0.00
	(0.67)	(0.28)	(0.00)	(0.00)		(0.59)	(0.58)	(0.00)	(0.00)
*T*R*	−220.39***	−23.56	−0.01	0.05	*T*R*	−90.82	−90.82	0.02	0.04
	(57.77)	(23.85)	(0.02)	(0.06)		(110.16)	(110.16)	(0.03)	(0.12)
VT	480.91	−376.39	0.02	−0.12	*VT*	165.12	165.12	0.02	−0.08
	(584.47)	(241.31)	(0.18)	(0.59)		(160.16)	(160.16)	(0.05)	(0.18)
VR	1.39	0.48***	−0.00	0.00	*VR*	−0.08*	−0.08*	0.00	0.00*
	(0.29)	(0.12)	(0.00)	(0.00)		(0.04)	(0.04)	(0.00)	(0.00)
Deg. of Freedom	14	14	14	14	Deg. of Freedom	19	19	19	19
R-squared	0.9299	0.8821	0.6346	0.3271	R-squared	0.7618	0.646	0.3284	0.4145
Adj R-squared	0.8598	0.7642	0.2693	−0.3459	Adj R-squared	0.5885	0.3885	−0.1601	−0.0114

Notes: Same as Table 10.2.

of the yield distributions for crops such as rice, sweet potatoes and bananas, but not for a perennial crop like the ponkan mandarin. The intra-annual variations for these two climate variables are also found to affect the mean yields of rice and sweet potatoes significantly, as well as the variances in banana and tomato yields. The climate variability may also widen the frequency distribution of yields for tomatoes and bananas, and thus a short sequence of years with low yields becomes more likely. If an inherent year-to-year instability in crop yields becomes a widespread phenomenon under future climate environments, worldwide efforts will be required to stabilize food supplies (Sinclair and Rawlins, 1993).

The results of this chapter also have implications for examining an agricultural crop insurance program, because the shape of yield distributions is one of the key parameters for designing crop insurance programs (Just and Weninger, 1999). If climate change shows great potential to alter the shape of crop yield distributions, we have to identify or estimate the direction and magnitude of these influences. Ignoring these influences will lead to distributional misspecifications, which in turn bias the calculation of insurance premiums and indemnity.

NOTES

1. The Mei-yu (or plum rain) season is an important season in Taiwan because the rainfall brought by the transition of weather systems is critical to the crops planted in early spring.
2. The data source is *Agricultural Yearbook* (1999). This estimate excludes the losses on soil erosion and farm facilities, but includes the damages from diseases and pests.

REFERENCES

Adams, R.M., B.A. McCarl, K. Segerson, C. Rosenzweig, K.J. Bryant, B.L. Dixon, R. Conner, R.E. Evenson and D. Ojima (1999), 'The Economic Effects of Climate Change on US Agriculture', in Mendelsohn, R. and J.E. Neumann (eds) (1999), *The Impact of Climate Change on the United States Economy*, Chapter 2, Cambridge: Cambridge University Press.

Ben Mechlia, N. and J.J. Carroll (1989a), 'Agroclimatic Modeling for the Simulation of Phenology, Yield and Quality of Crop Production: 1. Citrus Response Formulation', *International Journal of Biometeorology*, **33**, pp. 36–51.

Ben Mechlia, N. and J.J. Carroll (1989b), 'Agroclimatic Modeling for the Simulation of Phenology, Yield and Quality of Crop Production: 2. Citrus Model Implementation and Verification', *International Journal of Biometeorology*, **33**, pp. 52–65.

Chang, C.C. (2001), 'Potential Impact of Climate Change on Taiwan's Agriculture', *Agricultural Economics*, **27**, pp. 51–64.

Cheng, C.P. (1975), 'Natural and Technological Factors Contributing to Multiple-Crop Diversification in Taiwan', *The Philippine Economic Journal*, **14**, pp. 47–63.

Gallagher, P.W. (1986), 'US Corn Yield Capacity and Probability: Estimation and Forecasting with Nonsymmetric Disturbances', *North Central Journal of Agricultural Economics*, **8**, pp. 109–22.

Hoogenboom, G. (2000), 'Contribution of Agrometeorology to the Simulation of Crop Production and Its Applications', *Agricultural and Forest Meteorology*, **103**, pp. 137–57.

Hsu, H.H., Y.L. Chen and W.S. Kau (2001), 'Effects of Atmosphere–Ocean Interaction on the Interannual Variability of Winter Temperature in Taiwan and East Asia', *Climate Dynamics*, **17**, pp. 305–16.

IPCC (2001), *Summary for Policy Makers of the IPCC WG III Third Assessment Report*, Approved at 6th Session of WG III, Accra, Ghana.

Judge, G.G., R.C. Hill, W.E. Griffiths, H. Lütkepohl and T.C. Lee (eds) (1982), *Introduction to the Theory and Practice of Econometrics*, 2nd edn, John Wiley & Sons, New York.

Just, R.E. and Q. Weninger (1999), 'Are Crop Yields Normally Distributed?', *American Journal of Agricultural Economics*, **81**, pp. 287–304.

Kaufmann, R.K. and S.E. Snell (1997), 'A Biophysical Model of Corn Yield: Integrating Climatic and Social Determinants', *American Journal of Agricultural Economics*, **79**, pp. 178–90.

Kaylen M.S. and S.S. Koroma (1991), 'Trend, Weather Variables and the Distribution of US Corn Yields', *Review of Agricultural Economics*, **13** (2), 249–58.

Krishnamurti, T.N., R. Correa-Torres, M. Latif and G. Daughenbaugh (1998), 'The Impact of Current and Possibly Future Sea Surface Temperature Anomalies on the Frequencies of Atlantic Hurricanes', *Oceanographic Literature Review*, **45** (6), pp. 911.

Lam, W.F. (1996), 'Institutional Design of Public Agencies and Coproduction: A Study of Irrigation Associations in Taiwan', *World Development*, **24** (6), pp. 1039–54.

Lansigan, F.P., W.L. de los Santos and J.O. Coladilla (2000), 'Agronomic Impacts of Climate Variability on Rice Production in the Philippines', *Agriculture, Ecosystems and Environment*, **82**, pp. 129–37.

Luo, Y., P.S. Teng, N.G. Fabellar and D.O. Tebeest (1998), 'Risk Analysis of Yield Losses Caused by Rice Leaf Blast Associated with Temperature Changes Above and Below for Five Asian Countries', *Agriculture, Ecosystems and Environment*, **68** , pp. 197–205.

Matthews, R.B., M.J. Kropff, T. Horie and D. Bachelet (1997), 'Simulating the Impact of Climate Change on Rice Production in Asia and Evaluating Options for Adaption', *Agricultural Systems*, **54**, pp. 399–425.

Meejhl, G.A. and W.M. Washington (1996), 'El Niño-like Climate Change in a Model with Increased Atmospheric CO_2 Concentrations', *Oceanographic Literature Review*, **43** (12), pp. 1199.

Mendelsohn, R., W. Nordhaus and D. Shaw (1994), 'The Impact of Global Warming on Agriculture: A Ricardian Analysis', *American Economic Review*, **84**, pp. 753–71.

Mendelsohn, R., W. Nordhaus and D. Shaw (1996), 'Climate Impacts on Aggregate Farm Value: Accounting for Adaptation', *Agricultural and Forest Meteorology*, **80**, pp. 55–66.

Mendelsohn, R., W. Nordhaus and D. Shaw (1999), 'The Impacts of Climate Variation on US Agriculture', in R. Mendelsohn and J.E. Neumann (eds), *The Impact of Climate Change on the United States Economy*, Chapter 3, pp. 55–74, Cambridge: Cambridge University Press.

Park, W.I. and T.R. Sinclair (1993), 'Consequences of Climate and Crop Yield Limits on the Distribution of Corn Yields', *Review of Agricultural Economics*, **15** (3), pp. 483–93.

Rosenzweig, C., J. Philips, R. Goldberg, J. Carroll and T. Hodges (1996), 'Potential Impacts of Climate Change on Citrus and Potato Production in the US', *Agricultural Systems*, **52**, pp. 455–79.

Rosenzweig, C., M.L. Parry, G. Fischer and K. Frohberg (1993), *Climate Change and World Food Supply*, Research Report No. 3, Environmental Change Unit, Oxford: Oxford University.

Segerson, K. and B.L. Dixon (1999), 'Climate Change and Agriculture: The Role of Farmer Adaptation', in R. Mendelsohn and J.E. Neumann (eds), *The Impact of Climate Change on the United States Economy*, Chapter 4, Cambridge: Cambridge University Press.

Semenov, M.A. and J.R. Porter (1995), 'Climatic Variability and the Modelling of Crop Yields', *Agricultural and Forest Meteorology*, **73**, pp. 265–83.

Sinclair, T.R. and S.L. Rawlins (1993), 'Inter-Seasonal Variation in Soybean and Maize Yields under Global Environmental Change', *Agronomy Journal*, **85**, pp. 406–409.

Thompson, L.M. (1986), 'Climate Change, Weather Variability and Corn Production', *Agronomy Journal*, **78**, pp. 649–53.

Tobey, J., J. Reilly and S. Kane (1992), 'Economic Implications of Global Climate Change for World Agriculture', *Journal of Agricultural and Resource Economics*, **17**, pp. 195–204.

Zhang, R., A. Sumi and M. Kimoto (1996), 'Impact of El Niño on the East Asian Monsoon: A Diagnostic Study of the 86/87 and 91/92 Events', *Journal of Meteorology Society Japan*, **7**, pp. 49–62.

11. Will global warming cause heat stress?

Robert Mendelsohn and Daigee Shaw

1 INTRODUCTION

Many studies have found an empirical relationship between daily mortality in the elderly and heat waves. There are two competing theories which might explain the phenomenon: the threshold theory argues that risks increase whenever temperature exceeds a threshold and the adaptation theory hypothesizes that risks increase whenever temperature rises suddenly. The threshold theory predicts higher mortality in warmer climates whereas the adaptation theory predicts higher mortality with increased temperature variance. This study examines cross-sectional evidence between climate and mortality rates to test these two hypotheses. The data supports the adaptation theory finding higher elderly mortality in climates with higher temperature variance but not in places with warmer average temperatures.

With the increasing confidence that greenhouse gases will warm future climates (Houghton et al., 1996), there is growing concern about the future implications warming might have for society. One of the largest potential impacts concerns health (Pearce et al., 1996). There are several pathways linking climate and health, including changes in air pollution and vector-borne diseases (McMichael et al., 1996). In this chapter, we examine heat stress, another climate–health link (McMichael et al., 1996). There is compelling scientific evidence of heat stress from a set of intertemporal empirical studies which link unusually high summer temperatures with excess daily mortality rates in the elderly (Gover, 1938; Landsberg, 1981; Weihe, 1986; Kalkstein and Davis, 1989; Haines et al., 1993; Kalkstein, 1993; Kalkstein and Smoyer, 1993; Kunst et al., 1993; Pan et al., 1995). Periods of high temperature are consistently associated with higher daily mortality rates in many cooler temperate cities. Periods of unusually cold weather are also associated with higher daily mortality rates but this effect is less dramatic in the intertemporal record (McMichael et al., 1996).

There are two competing theories that can explain this observed association between temperature and elderly mortality: a threshold theory and an adaptation theory. The threshold theory, championed by Kalkstein and Davis

(1989), Kalkstein (1993) and Kalkstein and Smoyer (1993), argues that there exists a temperature threshold above which the elderly are at greater risk. This theory is consistent with the V-shaped relationships often found between elderly mortality and temperature (for example Kunst et al., 1993) at a single location. The hypothesis is that excess heat places added burdens on pulmonary and cardiac functions, causing increased failure rates. This hypothesis implies that one would observe more heat stress deaths in warmer places.

In contrast, the adaptation theory argues that heat stress is caused by rapid increases in accustomed temperature. Because it takes the body several days to make the required physiological adjustments (Kilbourne, 1992), sudden heat waves place individuals with little excess capacity at risk, namely the elderly. The theory explains why excess heat deaths occur in early summer with the first summer heat waves and not in late summer when temperatures are actually higher (Kalkstein, 1993). The adaptation theory also explains why the V-shaped relationship between mortality and temperature is observed to shift to the right as one examines populations living in warmer climates (Kunst et al., 1993; Pan et al., 1995; Honda et al., 1996). The cross-sectional data suggests that people make changes in their housing stock and behavior to adjust to average climate conditions mitigating the effect of average temperature. For example, people purchase more air conditioning in warmer climates, they exercise during cooler parts of the day and they stay out of the sun during the hottest times of the day. The adaptation theory suggests that locations with high temperature variance would result in more mortality but that heat stress mortality would not be higher in warmer climates.

Both theories are consistent with the intertemporal daily mortality record. However, the two theories have very different policy implications. First, the two theories suggest different health protection strategies. The threshold theory suggests precautions be instituted on all days which exceed a certain temperature, primarily in late summer when temperatures are highest. The adaptation theory implies that precautions should focus on sudden heat waves, primarily at the beginning of summer, when people are not yet used to the warmth. Second, the two theories have different implications for the relationship between global warming and health. The threshold theory predicts that global warming, by raising mean temperatures, would cause there to be many more days above the threshold. This has led some analysts to predict that global warming would substantially increase heat stress mortality (Kalkstein, 1993; McMichael et al., 1996), leading to 6600–9800 excess deaths in the United States alone. These deaths could be one of the largest harmful impacts of climate change, accounting for $6 to $37 billion of United States damages annually (Pearce et al., 1996). In contrast, the adaptation theory would argue that warmer temperatures would have minimal effect. Climate change would cause increased heat stress mortality only if it increased the variance of temperature (especially in the summer).

This study tests these two theories by examining a cross-section of all counties in the contiguous United States. County mortality rates for white elderly people are regressed on climate and other control variables. Climate normals for each season and interannual variation terms are both included. Separate analyses are performed for men and women. The regression coefficients reveal the impact of warmer average summer and winter temperatures and higher temperature variances on mortality. The cross-sectional evidence tests whether elderly mortality is affected by higher average temperatures or increased fluctuations in temperature.

The next section of the chapter reviews the methods and data used in the study. Section 3 presents the quantitative results. In section 4, we discuss the implications of the results for global warming policy.

2 METHODOLOGY AND DATA

Following Mendelsohn and Orcutt (1979), we adopt the following model of mortality rates for an individual:

$$D_i = BX_i + GZ_i + e_i \tag{11.1}$$

where D_i is the probability of death in a year for individual i, X_i is a vector of climate variables, Z_i is a vector of control variables which affect mortality rates and e_i is a random error term. Because mortality is rare, individual data is difficult to collect. We consequently rely on aggregate data collected at the county level. Because age, sex and race have profound effects on mortality rates, we analyse only the white population and conduct separate analyses by sex on only the elderly. Summing across individuals in a county yields the following aggregate model:

$$\sum D_i = B\sum X_i + G\sum Z_i + \sum e_i \tag{11.2}$$

Dividing by the population, n, for each age sex cell yields:

$$\bar{D} = B\bar{X} + G\bar{Z} + \bar{e} \tag{11.3}$$

where each variable is now the average measure for a county. The result is two (one for each sex) separate regressions of (11.3) with the unit of observation being counties (approximately 3000 observations) in the continental United States.

Because the independent socio-economic variables are measured as aggregates instead of as individual observations, aggregation error is introduced which will bias the estimated coefficients of the model towards zero. Further, there is limited information on other important confounding variables such as smoking, air pollution, diet, occupational exposure and exercise that could bias the results. Because of potential confounding omitted variables, cross-sectional studies must be interpreted cautiously. It is important to note, though, that the original cross-sectional results evaluating air pollution and mortality (Lave and Seskin, 1977; Mendelsohn and Orcutt, 1979) have largely been confirmed by subsequent longitudinal panel studies which have measured such variables (Dockery and Pope, 1994; Pope et al., 1995).

Mortality statistics are compiled from the National Center for Health Statistics from a complete census of all deaths for the lower 48 states. This study relies upon mortality rates measured over five years, from 1978 to 1982. The five years of data help focus the analysis on long-run effects rather than any phenomenon peculiar to a specific year. We do not take a longer series of data because the mortality data is linked with control variables measured in the 1980 US Census. The five years of data are averaged for a single long-term mortality rate. Mortality rates for white individuals were computed for men and women over 65 years of age.

Climatic data was obtained from the National Climatic Data Center, which gathers data from 5511 meteorological stations throughout the contiguous United States. The data include information on precipitation and temperature for each month from 1951 to 1980. We examine the climatologically normal variables – the 30-year average of each climatic variable for every station. The difference in long-run climate from place to place is relatively stable, picking up differences between the mid-continent versus the coast, the south versus the north and wet regions versus dry regions. Whether we used the 1951–80 or the 1961–90 climate normals would not change the estimates of climate across the data set very much. Of course, the same could not be said of daily weather, where it would make a great deal of difference. In this analysis, we collect data on normal monthly mean temperature and precipitation for January and July. We also include climate variation variables that measure how much climate has varied over the 30-year period. For each measure of climate, we have included the range for that variable over the 30-year period. In order to have climate variables for each county, we interpolate from the meteorological stations to the county centers. The interpolation is done using a sophisticated weighted regression technique (Mendelsohn et al., 1994). The means have been removed from the climate variables to facilitate interpretation of the regression coefficients.

Additional control variables for economic and social conditions come from the City and County Data Book. These control variables are not specific to age–sex cells but rather represent county averages facing all groups. The underlying hypothesis is that climate affects mortality. However, in an uncontrolled experiment such as this, it is important to control for other sources of variation in mortality rates. The basic regression introduces income, female labor force participation, farm employment, age, percentage divorced and the density of the population to control for these unwanted effects.

3 ANALYSIS

Several different specifications were tested to determine which sets of climate variables have an effect on human mortality. The t-statistics on the squared climate terms reveal that the climate variables have a quadratic relationship with mortality. Both climate normal and variation terms were generally statistically significant. Summer and winter variables are explored for both temperature and precipitation. The results shown in Table 11.1 thus include eight climate normal variables and four climate variation variables in each regression. The means have been removed from the climate variables to facilitate interpretation of the climate coefficients. Each linear coefficient represents the marginal effect of that variable evaluated at the mean. The quadratic term measures the curvature of the dose response function. Each observation is weighted by population in order to adjust for observed heteroscedasticity across counties.

The control variables in Table 11.1 reflect a number of non-climate mechanisms affecting mortality. Increased crowding and growth reduces mortality rates for the elderly, possibly reflecting access to better health care and social networks. Increased urbanization increases mortality, possibly capturing increased contact with contagious diseases amongst people. Increased density has mixed effects, possibly reflecting both influences above. Increased divorce rates and farms have mixed and minimal effects. Surprisingly, higher income is beneficial to women but harmful to men, perhaps reflecting ease of living on one hand and the stress of earning money on the other. Higher participation by women lowered the observed mortality rate. This effect could be due to the increased status and security that employment offers, the increased involvement that a job provides, or some other social characteristic that is spatially associated with places (metropolitan areas) where more women work. Higher average ages in the community are beneficial to the elderly, suggesting they do better in places where there are more elderly.

Table 11.1 Regression model of climate on elderly mortality

Climate variables	Male	Female	Control variables	Male	Female
JAN TEMP	−8.50	7.54	URBAN%	450.00	171.00
	(1.40)	(1.76)		(9.12)	(4.97)
JAN T SQ	−3.68	−4.94	CROWDING	−945.00	−1230.00
	(5.52)	(10.40)		(2.04)	(3.73)
JUL TEMP	−21.30	−65.80	DENSITY	13.30	−4.43
	(2.22)	(9.75)		(8.73)	(4.41)
JUL T SQ	3.92	5.46	POPGROWTH	−1340.00	−838.00
	(1.37)	(2.70)		(29.49)	(25.57)
JAN PREC	3.51	−1.71	DIVORCE	30.80	−24.60
	(4.50)	(3.17)		(0.63)	(0.72)
JAN P SQ	−0.020	−0.013	FEMLAB	−853.00	−2320.00
	(7.61)	(6.84)		(2.32)	(9.09)
JUL PREC	1.44	1.13	INCOME	−55.00	8.49
	(2.49)	(2.80)		(9.67)	(2.19)
JUL P SQ	0.004	0.002	FARM	−0.03	0.10
	(0.59)	(0.50)		(0.67)	(3.72)
JAN T VAR	19.90	−5.89	AGE	−1.22	−2.13
	(1.93)	(0.81)		(4.21)	(10.32)
JUL T VAR	80.60	49.30	CONSTANT	7310.00	5730.00
	(3.43)	(3.05)		(34.32)	(38.26)
JAN P VAR	−0.72	1.10			
	(2.46)	(5.48)			
JUL P VAR	0.40	0.18			
	(1.57)	(1.04)			
R^2	0.68	0.86			

Notes: The numbers in parenthesis are t-statistics. There are 3070 observations in each regression. Mortality rates are deaths per 100000. Observations weighted by population.

The focus of this study is upon the pattern of the summer temperature coefficients. The threshold theory suggests that the coefficients on mean temperature should be positive and significant whereas the adaptation theory suggests the coefficients on the variance of temperature should be positive and significant. In addition, because the adaptation theory also suggests the V-shaped relationship between mortality and temperature, we should observe negative coefficients for the linear summer temperature and positive coefficients for quadratic summer temperature. As is evident in Table 11.1, the coefficients of the linear summer temperature variable are negative and significant for both men and women and the coefficients of the quadratic summer temperature variable are positive for both men and women and are significant for women. There are lower elderly mortality rates in counties with higher average summer temperatures. The January temperature effects are mixed and not significant. These results are inconsistent with the threshold theory. In contrast, the summer coefficients of higher interannual temperature variation are positive and significant for both men and women. The empirical evidence thus provides strong support for the adaptation theory.

In order to get a sense of the magnitude of the predicted effects, we explore four climate scenarios. First, we explore the low, mid and high scenarios suggested by the IPCC (Houghton et al., 1996) that imply an average 1°C, 2°C and 3.5°C warming. We also include a scenario where the interannual variance of temperature increases by 20 per cent. Given the climate sensitivity functions estimated in Table 11.1, we calculate the expected number of excess deaths associated with each scenario. The results reported in Table 11.2 reveal that warming is expected to result in fewer deaths in the elderly. The mid-range effect of a 2°C warming would be to reduce elderly white deaths in the United States by 25000. This amounts to a 2 per cent reduction in average deaths (the average number of elderly deaths is 1190000). Although one may want to argue about whether or not warming will save lives, the evidence clearly does not support a prediction of increased mortality.

Table 11.2 also presents the results of increased variance. Increasing summer temperature variance would result in more deaths in the United States. The coefficients in Table 11.1 imply a 20 per cent increase in variance would lead to an additional 7200 deaths of both men and women (about a 0.7 per cent increase). This effect appears to be statistically significant. Increasing winter variance, however, has mixed effects, increasing mortality in women but decreasing mortality in men. Further, the effect is not statistically significant. Overall, the evidence supports the adaptation theory and rejects the threshold theory.

Table 11.2 Predicted excess elderly deaths caused by climate scenarios

Climate Scenario	Male	Female	Total
Uniform Warming			
1°C	−3454	−9336	−12790
	(1096)	(1137)	(1579)
2°C	−6864	−18527	−25391
	(2227)	(2310)	(3208)
3.5°C	−11897	−32043	−43940
	(4307)	(4476)	(6212)
20% Increase in Interannual Temperature Variability			
Summer	3756	3430	7186
	(1094)	(1126)	(1570)
Winter	2062	−908	1154
	(1067)	(1115)	(1543)

Notes: These estimates are for the white population in the contiguous United States over 65. The average annual deaths for elderly white males was 587000 over the sample period and for elderly white females, it was 603000. A negative number implies a reduction in deaths. The standard error is in parenthesis.

4 DISCUSSION

This study relies on cross-sectional analysis in order to help interpret daily mortality results. The consistent and widespread V-shaped relationship between daily mortality and climate is not under question. The issue is how to interpret the empirical relationship. There are two competing theories in the literature. One theory argues a temperature threshold. Mortality increases whenever temperatures rise above that threshold. A second theory argues that the issue is adaptation, and mortality increases only when temperature suddenly rises. Both theories explain the observed daily mortality phenomenon. Further, it is difficult to discern which theory is correct just using daily mortality data.

This study tests these two theories using cross-sectional analysis. If the threshold theory is correct, then one should see a relationship between annual mortality rates and higher average summer temperature. If the adaptation theory is correct, one should see a relationship between annual mortality and higher temperature variance. The regression results reveal that places with higher average summer temperatures have lower annual mortality rates in the elderly, which suggests that the threshold theory is not correct. The regression results also reveal that there is higher elderly mortality in places with higher summer interannual temperature variance. The results consequently support the adaptation theory of heat stress. The results explain casual observations such as why heat stress deaths are observed in cities such as Chicago (low average temperature and high variance) but not in places such as Miami (warm average temperatures and low variance) and why daily mortality rates are higher in the early summer than the late summer.

These results have implications for how society could adapt to the current climate. First, they suggest that heat waves in the beginning of summer are the most dangerous events to prepare for. Communities could adjust by providing special services for the elderly during these critical moments. For example, structures could be temporarily prepared which serve as cool shelter for the elderly during these events. Warnings could be provided which encourage the elderly to take special precautions during these episodes to reduce exposure to the heat and remain cool at nights.

The results also have implications for climate change impacts. Higher average temperatures are not predicted to cause an increase in overall mortality rates from heat stress. In fact, the results indicate that warmer summers would reduce mortality in the elderly in the United States. There is consequently no evidence suggesting that the predicted warming (rise in mean temperature) for the next century (2℃) would increase overall mortality. However, there is evidence that any increase in temperature variation would be harmful. At the moment, climate scientists are unwilling to predict that greenhouse gases will increase the variance of temperature (Houghton et al., 1996). Consequently, the current evidence does not suggest that increases in greenhouse gases would lead to increases in heat stress deaths. The empirical evidence suggests this link will exist only if greenhouse gases increase temperature variance.

APPENDIX

Table 11.A List of variables

Variable definition	Mean/(Std Dev)
JAN PREC Average monthly precipitation in January in mm	2.62/(1.87)
JUL PREC Average monthly precipitation in July in mm	3.62/(1.53)
JAN P SQ Square of JAN PREC	10.32/(15.58)
JUL P SQ Square of JUL PREC	15.48/(11.51)
JAN P VAR Range of JAN PREC from 1950 to 1980	5.9/(3.6)
JUL P VAR Range of JUL PREC from 1950 to 1980	8.4/(3.0)
JAN TEMP Average monthly temperature in January in ℃	31.6/(12.3)
JUL TEMP Average monthly temperature in July in ℃	75.8/(5.3)
JAN T SQ Square of JAN TEMP	1148/(790)
JUL T SQ Square of JUL TEMP	5770/(788)
JAN T VAR Range of JAN TEMP from 1950 to 1980	20.7/(3.6)
JUL T VAR Range of JUL TEMP from 1950 to 1980	8.5/(1.8)
AGE Median age in county in years	31.1/(3.9)
CROWDING Percentage of households with 1 or more people per room	048/(0.035)
DENSITY Ten thousand people per square mile	019/(0.156)
DIVORCE Percentage of males 15+ who are divorced	047/(0.023)
FARM Percentage of land in farmland	056/(0.031)
FEMLAB Percentage of adult females in labor force	400/(0.038)
INCOME Average income per capita in thousands of 1980 dollars	8.6/(2.0)
POPGROWTH Percentage change in county population from 1970 to 1980	166/(0.233)
URBAN% Percentage of county population in an urban area	355/(0.288)

REFERENCES

Dockery, D.W. and C.A. Pope (1994), 'Acute Respiratory Effects of Particulate Air Pollution', *Annual Review of Public Health*, **15**, pp. 107–32.

Gover, M. (1938), 'Mortality during periods of excessive temperature', *Public Health Reports*, **53**, pp. 1112–43.

Haines, A., P.R. Epstein and A.J. McMichael (1993), 'Global Health Watch: Monitoring the Impacts of Environmental Change', *The Lancet*, **342**, pp. 1464–69.

Honda, Y., M. Ono, A. Sasaki and I. Uchiyama (1996), 'Relationship between Ambient Temperature and Mortality Rate in Japan and Preliminary Projection of the Effect of Global Warming on Human Mortality', in I. Uchiyama and A. Sasaki (eds), *Proceeding of Health Effects and Risk Evaluation of Global Warming (1993–1995)*, pp. 27–33, Japan: National Institute of Public Health.

Houghton, J., L.G. Filho, N. Harris, A. Kattenberg and K. Maskell (eds) (1996), *Climate Change 1995: The Science of Climate Change*, Intergovernmental Panel on Climate Change (IPCC), Cambridge, UK: Cambridge University Press.

Kalkstein, L. (1993), 'Health and Climate Change: Direct Impacts on Cities', *The Lancet*, **342**, pp. 1397–99.

Kalkstein, L. and R. Davis (1989), 'Weather and Human Mortality: An Evaluation of Demographic and Interregional Responses in the United States', *Annals of the Association of American Geographers*, **79**, pp. 44–64.

Kalkstein, L. and K.E. Smoyer (1993), 'The Impact of Climate Change on Human Health: Some International Implications', *Experientia*, **49**, pp. 969–79.

Kilbourne, E.M. (1992), 'Illness Due to Thermal Extremes', in J.M. Last and R.B. Wallace (eds), *Public Health and Preventive Medicine*, Maxcy-Rosenau-Last, Norwalk, Appleton and Lange, 13th edn, pp. 491–501.

Kunst, A.E., C.W. Looman and J.P. Mackenbach (1993), 'Outdoor Air Temperature and Mortality in the Netherlands: a Time-Series Analysis', *American Journal of Epidemiology*, **137**, pp. 331–41.

Landsberg, H.E. (1981), *The Urban Climate*, New York: Academic Press.

Lave, L. and E. Seskin (1977), *Air Pollution and Human Health*, Baltimore: Johns Hopkins Press.

McMichael, A.J., A. Haines, R. Slooff and S. Kovats (eds) (1996), *Climate Change and Human Health*, Geneva: World Health Organization.

Mendelsohn, R. and G. Orcutt (1979), 'An Empirical Analysis of Air Pollution Dose-Response Curves', *Journal of Environmental Economics and Management*, **6**, pp. 85–106.

Mendelsohn, R., W. Nordhaus and D. Shaw (1994), 'Measuring the impact of global warming on agriculture', *American Economic Review*, **84**, pp. 753–71.

Pan, W., L. Li and M. Tsai (1995), 'Temperature extremes and mortality from coronary disease and cerebral infarction in elderly Chinese', *Lancet*, **345**, pp. 353–55.

Pearce, D., W. Cline, A. Achanta, S. Fankhauser, R. Pachauri, R. Tol and P. Vellinga (1996), 'The Social Costs of Climate Change: Greeenhouse Damage and the Benefits of Control', in J. Bruce, H. Lee and E. Haines (eds), *Climate Change 1995: Economic and Social Dimensions of Climate Change*, Intergovernmental Panel on Climate Change (IPCC), Cambridge, UK: Cambridge University Press.

Pope, C.A., D. Bates and M. Raizenne (1995), 'Health Effects of Particulate Air Pollution: Time for Reassessment', *Environmental Health Perspectives*, **133**, pp. 472–80.
Weihe, W.H. (1986), 'Life Expectancy in Tropical Climates and Urbanization', WMO Technical Conference Urban Climate, WMO-652, Geneva, Switzerland.

12. The impact of climate change on Asian Pacific countries

Robert Mendelsohn

1 INTRODUCTION

Using a new climate-impact model, this chapter examines how future climate change might affect the countries of Southeast Asia. The model combines detailed climate scenarios, sectoral features of each country, and climate-response functions to generate country-specific impacts by sector. Estimates are made for 1℃, 2℃ and 4.5℃ warming scenarios using two climate-response functions – estimated from experimental and cross-sectional data, respectively. The cross-sectional model predicts that most countries benefit from mild warming scenarios. The experimental model predicts that temperate countries may benefit but that tropical countries will likely suffer large damages from warming. Agriculture and non-market impacts explain most of the variation among scenarios and countries.

There have been extensive studies that have analysed the impacts of global warming (Bruce et al., 1996). Although there is a vast amount of information about the direct effects of warming on a host of resources (Watson et al., 1996), only a handful of studies have linked these direct effects to damages. The IPCC report relies heavily on expert opinion (Nordhaus, 1991; Cline, 1992; Titus, 1992; Tol, 1995 and Fankhauser, 1995) to make this leap. These expert judgements in turn rely largely on US studies (especially Smith and Tirpak, 1989). Extrapolation of these results to other countries using physical-impact measures – for example, miles of vulnerable coastline, Holdridge Life Zones, area of wetlands – require even more judgement (Fankhauser, 1995; Tol, 1995).

In this study, we take advantage of some recent research on impacts in the United States which estimates climate response functions (Mendelsohn and Schlesinger, 1999). These climate response functions are based on a set of comprehensive empirical studies measuring impacts to the American economy (Mendelsohn and Neumann, 1999). The climate response functions have been estimated for two different data sources. The first source relies on experimental data which is used to calibrate sophisticated simulation models of sectors. Welfare is estimated from various climate scenarios. A reduced form climate response function is then estimated from these results. The second approach relies on cross-sectional evidence of households, farms and firms across climate zones. This empirical approach leads directly to estimated climate response functions.

The information about market impacts described above is supplemented by a hedonic wage study which measures the non-market impacts of climate (Mendelsohn, 2001). This study examines wages across climate zones and calculates whether people want a premium to move to a warmer climate or whether they will accept a lower wage in order to move there. The study measures the non-market aspects of climate that are associated with choosing a county to live in, such as the weather, the ecosystem that accompanies the climate, outdoor recreation opportunities, energy costs and known links with health. The study does not capture non-use values such as preferences for the protection of remote endangered species or preferences about global ecological outcomes. The study finds that Americans want to move to warmer climates but that there is a limit on how warm they want the climate to get. Countries in the tropics, for instance, are already too hot according to this American study and any further warming would be deleterious.

Combining detailed climate projections from climate models, detailed information about each sector in each country and these climate response functions, one can estimate the impacts for every country in the world (Mendelsohn et al., 2000b). This chapter focuses on the impacts on Southeast Asian countries. The distribution of predicted impacts will vary across a host of climate models (Mendelsohn et al., 2000a) because some models predict a lot of warming in the tropics and others do not. This chapter does not explore the uncertainty in climate predictions. Instead, we examine three different climate change scenarios for each country, a 1℃, 2℃ and 4.5℃ warming with a 7 per cent increase in precipitation. In practice, climate models predict warming will be greater towards the poles than near the equator so that northern countries like Japan are more likely to experience greater warming than tropical nations.

2 METHODOLOGY

In this section, we describe the future world scenarios used and the three components of our modeling effort: the climate model, the sectoral data and the climate-response functions. The future world scenarios were taken from Houghton et al. (1994, 1996) and describe the low, mid and high scenarios used by the IPCC. The scenarios include emission trajectories for carbon dioxide, methane, nitrous oxide and chlorofluorocarbons and predicted temperature and sea-level changes by 2100. Assuming that the emission scenarios vary largely because of the size of the world economy, we also calculate alternative economic growth rates for each scenario. The low emission scenario (IS92c) is associated with a 0.8 per cent annual economic growth rate. By 2100, this will result in a CO_2 concentration of 500 ppmv (parts per million volume), a 1℃ increase in global temperature, a 3.5 per cent precipitation increase and a 15cm increase in sea level. The mid scenario (IS92a) is associated with a 1.9 per cent annual economic growth rate and in 2100, a CO_2 concentration of 650 ppmv, a 2℃ increase in global temperature, a 7 per cent increase in precipitation and a 50cm increase in sea level. Finally, the high scenario (IS92e) is associated with a 2.6 per cent annual economic growth rate, and in 2100, a CO_2 concentration of 950 ppmv, a 4.5℃ increase in global temperature, a 15 per cent increase in precipitation and a 100cm increase in sea level. These scenarios do not bracket all possible future outcomes over the next century, because the high temperature scenarios are associated with high greenhouse-gas levels and high economic development. The scenarios nonetheless provide a wide and internally consistent range of forecasts.

2.1 Climate

The climate modeling determines the change in annual surface–air temperature and prccipitation for 177 countries using the geographical distributions simulated by two versions of the University of Illinois at Urbana-Champaign (UIUC) atmospheric general circulation/mixed-layer-ocean model using an 11-layer model between the earth's surface and 50 mb (UIUC11) (Schlesinger et al., 1997a), with a horizontal resolution of 4° latitude × 5° longitude. This UIUC11 experiment involves two simulations: control (345 ppmv, $1 \times CO_2$) and doubled ($2 \times CO_2$) CO_2 concentrations (Schlesinger et al., 1997b). Country-specific normalized surface–air temperature and precipitation changes were calculated by averaging grid-cell values within national borders.

2.2 Sectoral Data

The sectoral data incorporates country-specific information including GDP, average land value, population, cropland, forestland and coastline. The size of each economic sector is measured for each country. We project growth in each sector so that we can measure future sensitivity. For example, given the historic reduction of agriculture as a fraction of GDP, we project that future agriculture sectors will grow more slowly than GDP as a whole. GDP projections suggest that developing countries will grow more rapidly than OECD countries (Houghton et al., 1994).

2.3 Response Functions

The response functions to climate change are based on empirical studies that have been carefully designed to include adaptation by firms and people to climate change. Separate response functions are estimated for agriculture, forestry, coastal resources, commercial and residential energy, and water. We use two alternative response functions for most sectors to demonstrate the sensitivity of our results to the type of response function used. The origins of these response functions are carefully developed in Mendelsohn and Schlesinger (1999) and Mendelsohn (2001).

The economic response functions are based on a collection of sectoral studies for the United States completed by a team of leading impact experts (Mendelsohn and Neumann, 1999). These studies use a variety of empirical approaches including agronomic–economic models and ecological–forestry models to build consistent, comprehensive estimates of damages in each sector. Using the net results from each sector, we have constructed a reduced-form model that links climate scenarios and welfare impacts for each sector to temperature and precipitation (Table 12.1).

We also use a set of Ricardian studies that relies on cross-sectional comparisons to reveal how agriculture, forestry and energy would respond to climate change (Mendelsohn and Schlesinger, 1999). The agricultural Ricardian model (Mendelsohn et al., 1994) measures how long-term farm profitability varies with local climate, controlling for other factors. The forestry model is based on a similar cross-sectional analysis of the effect of climate on the present value of timber grown in the United States. The energy study relies on an analysis of energy expenditures in the commercial and residential sectors.

These studies of market impacts are supplemented by the non-market study of wages (Mendelsohn, 2001). Although the hedonic wage study provides an estimate of these effects, the non-market impacts are far more uncertain than the market estimates. Readers should interpret the non-market effects with caution.

Most of the response functions imply that the net productivity of sensitive economic sectors is a quadratic function of temperature (Mendelsohn and Schlesinger, 1999). *Ceteris paribus*, starting from cool temperatures, each economic activity increases in value as temperature increases to some maximum value, and then decreases as temperature increases beyond that point. As verification, these response functions are consistent with what we know about global economic productivity, where the most profitable sites for most climate-sensitive activities lie in the temperate or sub-tropical zones.

The response functions were all calibrated for the United States. To extrapolate from the US to other countries, we assume that: (1) agricultural and forestry responses are proportional to cropland and forestland, respectively; (2) coastal damages from sea-level rise are proportional to the amount of coastline and the average value of land, the latter approximated using GDP divided by area; (3) energy impacts are proportional to GDP; and (4) water impacts are proportional to agricultural GDP. We also assume that the level of development does not affect climate sensitivity, which is a questionable assumption. By ignoring this development effect, the study has probably underestimated the damages in tropical countries from warming. These assumptions can be relaxed when systematic country-specific impact studies are completed.

Table 12.1 Aggregate net impacts in 2100 on Southeast Asian countries from alternative climate change scenarios

	Climate Scenarios					
	1℃		2℃		4.5℃	
Models	EXP	CRS	EXP	CRS	EXP	CRS
China	20.5	4.3	83.3	18.7	237.6	54.6
Japan	2.4	0.5	13.9	5.4	27.8	19.2
N. Korea	−0.1	0.2	0.5	0.9	0.9	2.8
New Zealand	0.3	0.1	1.3	0.4	2.2	1.1
Russia	67.8	14.1	268.2	48.2	886.1	144.8
S. Korea	0.5	0.3	2.2	1.5	1.8	5.3
Australia	−5.0	1.5	−21.1	4.4	−77.4	8.0
Singapore	−0.7	−0.2	−3.5	−0.8	−16.8	−3.5
Taiwan	−0.7	−0.1	−3.9	−0.4	−21.3	−2.0
Bangladesh	−0.3	0.1	−3.0	−1.1	−19.4	−9.9
Cambodia	−0.3	0.2	−0.6	0.7	−1.1	1.3
India	−19.0	16.6	−90.8	46.4	−443.2	92.4
Indonesia	4.1	−0.7	7.0	−9.9	−7.9	−64.5
Laos	0.4	0.0	−0.7	0.0	−1.5	−0.2
Malaysia	0.7	−0.5	0.4	−3.6	−8.1	−19.4
Myanmar	−1.3	0.5	−5.8	1.2	−27.5	1.0
Philippines	0.3	0.1	−1.0	−1.3	−13.0	−11.1
Thailand	−3.5	1.2	−16.1	2.6	−73.4	1.4
Vietnam	−1.4	0.3	−5.5	0.9	−22.5	1.7

Note: Welfare estimates measured in billions of 1990 USD using cross-sectional (CRS) and experimental (EXP) models.

3 RESULTS

In this chapter we explore the annual impacts which would occur in 2100 under three forecasts of future world conditions using the low, mid and high scenarios presented by the IPCC (Houghton et al., 1996). For each scenario, we apply UIUC11 to predict future world climate. The faster world growth is, the greater the emissions and thus the higher the global warming by 2100. For each climate scenario, we apply two empirically based climate-response functions, the reduced-form model and the Ricardian model (Mendelsohn and Schlesinger, 1999).

The impact model calculates climate impacts on countries as we project them to appear in the future. This introduces another source of uncertainty because it is difficult to predict what economic conditions will be like in a century's time for each sector and each country. However, climate change will take many decades to unfold and so impacts must be evaluated in terms of future conditions. To keep impacts in perspective, it is important to recognize that, according to the mid scenario, the world economy is predicted to increase from $21 trillion today to $172 trillion by 2100 (all financial estimates are in 1990 US$). The world economy is projected to increase to $51 trillion in 2100 with the low-emission scenario and to $372 trillion with the high-emission scenario.

Table 12.1 presents the aggregate Southeast Asian country results in billions of 1990 USD for all 3 climate scenarios. Countries that currently are warm are more likely to be harmed by warming. The magnitude of the damages increases with the temperature change. Both the 1℃ and 2℃ warming scenarios generally entail only small benefits or damages. However, several countries in warmer climates are severely affected by warming in the neighborhood of 4.5℃. With the experimental response function, India takes a large loss from extreme warming. Australia and Thailand also take large losses, with countries like Myanmar and Vietnam suffering large effects relative to their GDP. Because warming has the largest effect on agriculture, countries with large agricultural sectors, such as Australia, India and China, are affected the most.

Countries that are currently in the temperate zone will benefit from warming. For all levels of warming and both climate response functions, China, Japan, North Korea and South Korea all benefit from global warming. China is predicted to have the largest gains from warming as much of their cooler north and interior become more productive.

The response function makes an enormous difference for some countries. India and Australia suffer large damages from climate change according to the experimental climate response function but gain according to the cross-sectional response function. For most other countries, the experimental results are the same sign but more extreme than the cross-sectional results. That is gains are larger and damages are larger with the experimental model. The only exception to this rule concerns Indonesia and Malaysia, which are expected to have a large increase in precipitation as temperature rises. The large precipitation increases lead to large projected damages in the cross-sectional forecasts for both countries, which exceed the projected experimental damages.

Severe global warming scenarios can cause billions of dollars of damages or benefits in the region. Although these effects are large in absolute size, one must remember that they will occur in 2100. The overall magnitude of the market impacts is relatively small with respect to the expected GDP for each country by then. The poorest countries suffer the greatest damages as a fraction of GDP. Most of the poorest nations in Southeast Asia would lose between 2 and 3.8 per cent of GDP with the more severe experimental results and the severe 4.5℃ scenario. Only Bangladesh has lower damages of 1.7 per cent. The moderate-income countries suffer only modest damages in this scenario between 0.1 per cent and 1.1 per cent, with the exception of Australia (1.8 per cent) and Thailand (2.0 per cent). The severe climate scenario with a severe climate response function would seriously affect several countries in the region. However, all other scenarios would likely result in damages of less than 1 per cent of GDP.

The effects in Southeast Asia resemble effects found throughout the world. The more temperate countries in the world are likely to benefit from warming. The tropical countries, because they are already hot, are more likely to be damaged. Agriculture is especially sensitive to warming and explains most of the variation in effects across countries. Non-market effects are also important for relatively large and prosperous countries. Only Japan suffers high coastal damages largely because of relatively extensive coasts and very high valued land.

Tables 12.2 and 12.3 present how the sectoral responses differ between the two climate-response functions. In both tables we rely on the middle climate scenario of 2℃ warming. Table 12.2 displays the cross-sectional results and Table 12.3 the experimental results. For this warming scenario, agriculture has the largest effects and it explains most of the variation across countries. In Table 12.2, the benefits or damages in the agricultural sector explain almost 100 per cent of the overall net effects. If there were no agricultural impacts, the net effects in Table 12.2 would be a small fraction of GDP (about 0.05 per cent). In Table 12.3, non-market effects are also important, although not as large as agriculture.

Agriculture explains most of the variation in impacts across scenarios and countries. The countries with the largest agricultural effects tend to have large amounts of cropland. Agricultural benefits are concentrated in high-latitude countries whereas damages tend to increase as one nears the equator. China has the greatest benefits because of large areas of farmland and its temperate climate. India and Australia have either large agricultural losses or gains depending upon the response function chosen.

Damages from increased energy costs for heating and cooling explain some of the differences amongst scenarios. In the mid-scenario, energy effects are projected to be damaging for all the tropical countries but mildly beneficial to the temperate countries. These damages are projected to be concentrated in the tropics as people struggle to cope with high heat. The expense of cooling outweighs the savings in heating in these countries.

Table 12.2 Sectoral climate impacts (cross-sectional)

(Billions USD)

Country	Agr	Ene	For	Wat	Cst	Non-market
China	15.5	0.5	1.1	−4.5	−0.1	6.0
Japan	0.5	−0.8	0.2	−0.0	−2.2	7.8
N. Zealand	0.4	0.0	0.0	−0.1	0.0	0.1
N. Korea	0.4	0.2	0.0	−0.0	−0.2	0.6
Russia	17.8	15.4	11.0	−9.1	−0.1	13.1
S. Korea	0.4	−0.1	0.0	−0.0	−0.2	1.4
Australia	5.8	−1.3	0.1	0.0	−0.1	−0.3
Singapore	0.0	−0.3	0.0	0.0	−0.4	−0.2
Taiwan	0.0	−0.3	0.0	0.0	−0.2	0.2
Bangladesh	−1.7	−0.2	0.0	1.0	−0.0	−0.1
Cambodia	0.6	−0.0	−0.0	0.1	0.0	0.0
India	41.3	−2.0	0.1	1.7	−0.2	−1.6
Indonesia	−13.5	−0.6	0.5	4.4	−0.2	−0.5
Laos	0.1	−0.0	0.0	−0.1	0.0	0.0
Malaysia	−4.3	−0.3	0.1	1.0	−0.0	−0.2
Myanmar	1.0	−0.1	0.0	0.3	−0.0	−0.1
Philippines	−1.8	−0.2	0.0	1.0	−0.1	0.2
Thailand	2.6	−0.5	0.0	1.0	−0.0	−0.4
Vietnam	1.1	−0.1	0.0	−0.1	−0.0	0.9

Note: Welfare measured in 2100 with 2°C increase

The remaining market effects are from water, forests and coasts. In the middle scenario, the water system effects are expected largely to gain from increased runoff, although a few countries suffer water reductions. For most countries, there is an increase in precipitation which exceeds the increase in evapotranspiration, increasing runoff. This leads to benefits primarily in the agricultural sector from increased irrigation. Forestry is projected to enjoy a gain because of warming. The carbon fertilization and increased temperatures are expected to increase forest productivity and expand productive biomes (Haxeltine and Prentice, 1996). Coastal damage from sea-level rise is projected in every scenario. However, the magnitude of coastal damages is predicted to be much smaller than earlier studies thought, because of intertemporal adaptation. That is, we expect that protection will occur only as needed allowing these costs to be postponed until late in the century. We also assume that coasts will migrate inward so that coastal boundaries will not be lost, but rather interior land. Japan suffers the greatest damage from sea-level rise because of the extraordinary value of Japanese land.

Table 12.3 Sectoral climate impacts (experimental)

(Billions USD)

Country	Agr	Ene	For	Wat	Cst	Non-market
China	63.5	1.6	0.5	−4.5	−0.1	22.2
Japan	1.1	0.1	0.0	−2.2	14.7	0.2
N. Zealand	0.8	0.0	0.0	−0.1	0.0	0.1
N. Korea	−1.7	0.2	0.0	−0.0	−0.2	2.3
Russia	193.0	4.5	6.7	−9.1	−0.1	73.3
S. Korea	0.8	−0.1	0.0	0.0	−0.2	1.7
Australia	−14.9	−0.7	0.5	−0.0	−0.1	−6.0
Singapore	0.0	−0.3	0.0	0.0	−0.4	−2.8
Taiwan	0.2	−0.5	0.0	0.0	−0.2	−3.9
Bangladesh	−1.6	−0.3	0.0	1.0	−0.0	−2.2
Cambodia	−0.7	−0.0	0.1	0.1	0.0	−0.1
India	−68.6	−3.4	0.6	8.7	−0.1	−28.1
Indonesia	10.0	−1.0	1.6	4.4	−0.2	−7.9
Laos	−0.6	−0.0	0.1	−0.1	0.0	−0.1
Malaysia	2.7	−0.4	0.3	1.0	−0.0	−3.2
Myanmar	−4.7	−0.2	0.2	0.3	−0.0	−1.5
Philippines	1.3	−0.4	0.1	1.0	−0.1	−2.9
Thailand	−10.0	−0.7	0.1	1.0	−0.0	−6.4
Vietnam	−4.3	−0.1	0.0	0.1	−0.0	−1.0

Note: Welfare measured in 2100 with 2°C increase

The final important effect is the non-market effect from warming. The tropical countries are all expected to prefer their current climate to any warmer future. The magnitude of damages depends on per capita income and the size of the population. The temperate countries are expected to prefer a slightly warmer climate. The non-market effects are beneficial for the temperate countries.

4 CONCLUSION

This study has several limitations: (1) the response functions were calibrated only for the United States; (2) the sectoral information about each country is limited; (3) non-market effects are included in only a coarse manner; (4) the resolution of the GCM is coarse relative to the size of small countries; (5) the transient response of the climate system is not considered; (6) the simulated climate changes are due only to increased CO_2 and not to the partially compensating effects of anthropogenic sulfate aerosols. Nevertheless, the results of this study show that it is possible to integrate detailed climate modeling and empirical climate-response functions to generate country -specific impact estimates.

The results suggest that the impacts of climate change are not likely to be uniform from country to country. Although market effects are expected to be relatively small as a fraction of GDP, there are important differences between what is expected to happen to countries in different climate zones. Countries who currently enjoy a temperate climate will fare relatively well with warming. Countries in tropical climates could well be hurt by even a small amount of warming. All of these quantitative estimates are uncertain. Continued research will undoubtedly improve these estimates, especially concerning non-market effects and effects in developing countries. Nonetheless, these country-specific impacts are informative and provide countries with our best currently available estimates. Future negotiations on international agreements to control greenhouse gases should take these country-specific responses into consideration.

REFERENCE

Bruce, J., H. Lee and E. Haites (eds) (1996), *Climate Change 1995: Economic and Social Dimensions of Climate Change, IPCC Report*, Cambridge: Cambridge University Press.

Cline, W. (1992), *The Economics of Global Warming*, Washington, DC: Institute for International Economics.

Fankhauser, S. (1995), *Valuing Climate Change: The Economics of the Greenhouse*, London: Earthscan.

Haxeltine, Alex and I. Colin Prentice (1996), 'BIOME3: An Equilibrium Terrestrial Biosphere Model Based on Ecophysiological Constraints, Resource Availability, and Competition Among Plant Functional Types', *Global Biogeochemical Cycles*, **10** (4), pp. 693–709.

Houghton, J.T., L.G. Meira Filho, J. Bruce, H. Lee, B.A. Callander, E. Haites, N. Harris and K. Maskell (eds) (1994), *Climate Change 1994: Radiative Forcing of Climate Change and An Evaluation of the IPCC IS92 Emission Scenarios*, Cambridge: Cambridge University Press.

Houghton, J.T., L.G. Meira Filho, B.A. Callander, N. Harris, A. Kattenberg and K. Maskell (eds) (1996), *Climate Change 1995: The Science of Climate Change*, Cambridge: Cambridge University Press.

Mendelsohn, R. (2001), 'The NonMarket Impacts of Global Warming: A Hedonic Wage Study', in D. Maddison (ed.) (2001), *The Amenity Value of the Global Climate*, Ottawa: Renouf Publishing.

Mendelsohn, R. and J. Neumann (eds) (1999), *The Economic Impact of Climate Change on the US Economy*, Cambridge: Cambridge University Press.

Mendelsohn, R. and M. Schlesinger (1999), 'Climate Response Functions', *Ambio*, **28**, pp. 362–66.

Mendelsohn, R., W. Nordhaus and D. Shaw (1994), 'The Impact of Global Warming on Agriculture: A Ricardian Analysis', *American Economic Review*, **84**, pp. 753–71.

Mendelsohn, R., M. Schlesinger and L. Williams (2000a), 'Comparing Impacts Across Climate Models', *Integrated Assessment*, **1**, pp. 37–48.

Mendelsohn, R., W. Morrison, M. Schlesinger and N. Adronova (2000b), 'Country-Specific Market Impacts from Climate Change', *Climatic Change*, **45**, pp. 553–69.

Nordhaus, W. (1991), 'To slow or not to slow: The economics of the greenhouse effect', *Economic Journal*, **101**, pp. 920–37.

Schlesinger, M.E., N.G. Andronova, B. Entwistle, A. Ghanem, N. Ramankutty, W. Wang and F. Yang (1997a), 'Modeling and Simulation of Climate and Climate Change, Past and Present Variability of the Solar-Terrestrial System: Measurements, Data Analysis and Theoretical Models', in *Proceedings of the International School of Physics 'Enrico Fermi'*, 25 June–5 July 1996, Varrena, Italy.

Schlesinger, M.E., N. Andronova, A. Ghanem, S. Malyshev, T. Reichler, E. Rozanov, W. Wang and F. Yang (1997b), 'Geographical Scenarios of Greenhouse-Gas and Anthropogenic-Sulfate-Aerosol Induced Climate Changes', Climate Research Group, Department of Atmospheric Sciences, University of Illinois at Urbana-Champaign, Urbana, IL.

Smith, J. and D. Tirpak (1989), *The Potential Effects of Global Climate Change on the United States: Report to Congress*, Washington DC: Environmental Protection Agency.

Titus, J.G. (1992), 'The Cost of Climate Change to the United States', in S.K. Majumdar, L.S. Kalkstein, B. Yarnal, E.W. Miller and L.M. Rosenfeld (eds), *Global Climate Change: Implications, Challenges and Mitigation Measures*, Pennsylvania Academy of Sciences, Easton, PA.

Tol, R.S.J. (1995), 'The damages costs of climate change: Towards more comprehensive calculations', *Environmental and Resource Economics*, **5**, pp. 353–74.

Watson, R., M. Zinyowera, R. Moss and D. Dokken (eds) (1996), *Climate Change 1995: Impacts, Adaptations and Mitigation of Climate Change, Scientific Technical Analyses*, Cambridge: Cambridge University Press.

PART IV

Policy Instruments

13. Reducing cost uncertainty and encouraging ratification of the Kyoto Protocol[1]

Raymond Kopp, Richard Morgenstern, William Pizer and Frédéric Ghersi

1 INTRODUCTION

Put into force, the Kyoto Protocol represents a significant commitment to the goal of stabilizing atmospheric concentrations of greenhouse gases. Unlike the earlier United Nations Framework Convention on Climate Change (UNFCCC) which called for voluntary efforts to reduce greenhouse gas (GHG) emissions, the Kyoto Protocol establishes binding emission limits on participating countries. This shift from voluntary to mandatory action reflects a substantial elevation of the global commitment to mitigate the threat of climate change.

While the movement from voluntary action to firm commitments is the hallmark of the Kyoto Protocol, the adoption of commitments does not necessarily imply there is a need to adopt quantity targets, as called for by Article 3 of the protocol.[2] Committing to reduce GHG emissions can be accomplished by means other than the imposition of binding quantity targets. For example, the implementation of taxes on GHG emissions (most notably carbon dioxide emissions) is the obvious alternative to binding quantity targets. In a world where the cost of GHG abatement is known, any desired level of emissions can be attained with a tax mechanism (termed a 'price instrument').

Economists are often asked if quantity instruments (binding quantity targets) and price instruments (for example, emission taxes) yield the same environmental result; is there any reason to prefer one over the other? In general, if the cost of limiting emissions is known with certainty and the benefits of reduced emissions are similarly known, the price and quantity approaches are perfect policy substitutes. However, as Weitzman (1974) pointed out a quarter of a century ago, when abatement costs are uncertain, as in the case of GHGs, the situation can be quite different. On the one hand, quantity restrictions can be beneficial when incremental damages increase rapidly with the level of emissions or when marginal costs are relatively flat and predictable. In that case, quantity restrictions are preferred because they prevent emissions from rising above a 'safe' level and do not risk cost surprises. On the other hand, when health or environmental damages are not very sensitive to short-term emission levels, or when concerns exist about potentially high costs, the undesirable side effects of quantity restrictions may dominate. In that case, price-based instruments are preferred.

Which of these situations applies to greenhouse gases? The first observation is that GHGs represent a canonical example of what is known as a 'stock pollutant', one in which the damages are a function of total accumulation in the environment and annual emission flows are a relatively small part of the total stock. Thus, there is little basis to believe that *short-term* increases in emissions cause large environmental damages.

What about the longer-term damages as concentrations rise over time? A great deal of attention has been given to the possibility of some extreme event (like a dramatic rise in the sea level from melting polar ice caps or a shift in the Gulf Stream which destabilizes global temperature and weather). The most likely scenario is that climate change will occur gradually in response to growing atmospheric concentrations of greenhouse gases and most evaluations of potential consequences of climate change are based upon this assumption. However, the potential clearly exists – with unknown probability – for disproportionately large responses to even small disturbances in the climate system.

Pizer, in a series of papers (1996, 1998, 1999, 2001, 2002), used a world-scale, computable general equilibrium model, capable of explicitly incorporating uncertainty, to demonstrate that price instruments are preferable on social welfare grounds to quantity targets for the abatement of GHG emissions. Following from Weitzman that relatively flat marginal benefits/damages favor taxes, Pizer finds that an optimal tax designed to control GHGs generates gains five times greater than the optimal permit policy. As a result of his model simulations, Pizer (2002) finds that the optimal permit policy leads to lower welfare gains than a tax instrument.

Pizer's results suggest that the binding quantity targets of the Kyoto Protocol are not the socially optimal policy mechanism with which to implement 'commitments' under the UNFCCC. However, it is likely that few world leaders are familiar with Weitzman's theoretical results or Pizer's empirical work. And even if they were, the politics of global climate policy would, in their view, trump the ideas put forth by a band of economists. Moreover, in the unlikely case these leaders were persuaded by the economists, they would argue that the negotiation of the protocol has gone too far to think about abandoning the quantity targets and trying to develop and negotiate a new price-based approach.

It is, of course, possible to combine price and quantity policies in a hybrid approach that establishes binding emissions targets as long as costs remain reasonable, and to allow the target to rise somewhat if costs are unexpectedly high. What is needed is a penalty, specified in advance and paid by the source in case its emissions exceed the quantity restrictions set for that source. One way to achieve this goal is to establish a penalty per unit of emissions in excess of the quantity restriction. Such a hybrid policy (1) fixes emissions targets that are binding as long as costs remain reasonable and (2) allows the target to rise somewhat if costs are unexpectedly high. In practical terms, the hybrid, or 'safety valve', would involve an initial allocation of permits followed by the subsequent sale of additional permits to be made available at a fixed trigger price.

The purposes of this chapter are threefold. First, we set aside the economic social welfare arguments and argue there are significant political reasons to advocate a switch from a quantity-based to a price-based GHG abatement approach. Moreover, we argue that the quantity target imposes political difficulties so great that, unless they are addressed, the protocol may not enter into force, or might enter into force, but by a slim margin (probably without the United States). Second, we show how a movement from a quantity-based approach to a price-based approach can be effected with only simple alterations to the protocol, thereby avoiding the need to re-open the Kyoto process and renegotiate the targets, as some have suggested. Finally, we implement our proposal in the context of a simulation model and examine the effect of the proposal on global GHG emissions.

2 THE POLITICAL DIFFICULTY OF QUANTITY TARGETS

The difficulty raised by quantity targets begins and ends with compliance. As the protocol currently reads, a country would be in compliance with the protocol if its GHG emissions were at or below its assigned amount during the compliance period, or if it purchased assigned amounts of Clean Development Mechanism (CDM) credits on the world market to cover any shortfall. Assuming some meaningful penalties for non-compliance, a country will ratify the protocol only if it believes it can comply with its assigned quantity target.

But governments face vast uncertainty as they attempt to develop and implement polices and measures intended to meet their Kyoto commitments. This uncertainty involves both the likelihood of meeting a quantity target with a given set of policies and the political and economic cost of meeting the target. We argue that the presence of these uncertainties is a major political stumbling block to ratification of the protocol.

Consider a country that wishes to use a comprehensive carbon tax to meet its Kyoto commitment. The country employs the best energy–economy models at its disposal to choose the level of a carbon tax to meet the emission target. The country's policy makers then succeed in clearing all the political hurdles needed to pass the tax into law. However, to the country's great dismay, it finds out years later (during its commitment period) that the tax was too low, due perhaps to incorrect baseline assumptions or some other modeling inadequacies. This country acted in good faith, made politically difficult decisions and paid the economic cost of the tax, now only to be told that it is not in compliance with its commitments and thus subject to sanctions.

What went wrong? The answer is that a country choosing a tax-based approach will always be uncertain about its ability to meet a quantitative emission target owing to the reality of less-than-perfect modeling and economic forecasting. The situation is even worse for a country that chooses to use a mix of policies and measures, which might include taxes, tradable permits and prescriptive measures like efficiency standards. In this case, the difficulty of modeling any mix of these approaches is greater than modeling the tax alone, potentially giving rise to even larger errors.

The dictates of politics will generally lead countries to choose policies and measures that aim to keep economic costs as low as possible. This political incentive suggests that most forecasting errors (owing to incorrect economic assumptions or models) will lead to an undershooting of abatement targets. If a country wanted to reduce its compliance uncertainty, it could do so only by building in a margin of safety (for example, legislating a carbon tax 50 per cent higher than that predicted by the models), thereby raising expected costs. However, all things equal, the greater the expected cost, the less likely is a country to ratify the protocol.

Even if countries were willing to undertake high levels of effort, in terms of political and economic cost, to abate GHG emissions, not knowing if that level of effort would be 'good enough' in terms of Kyoto compliance will deter these countries from taking on commitments. Conceptually, a country could impose a fully upstream carbon permit system and meet its Kyoto emission target with certainty. While such an approach hits the compliance target, it does so at the expense of wholly uncertain GHG control cost. This situation is the exact opposite of the problem discussed above when attempting to set the right carbon tax. In the permit case, you know the emission level, but you can't know with certainty the cost. In the tax case, you know the cost with certainty (in terms of the tax), but you can't know the emission reductions that would result.

The binding-quantity emission requirements make the cost of participation in the Kyoto Protocol difficult to estimate and potentially quite high. Unexpected economic growth, the availability of alternative fuels, the development of new technologies and the details of international implementation all lead to uncertain estimates for both the required reductions and the cost of those reductions in individual countries. Uncertainty about these costs has led many countries to question both whether they should ratify the agreement and, once they do, whether all of the participating countries will be able to meet their targets by the end of the commitment period in 2012.

3 TRANSFORMING A QUANTITY TARGET TO A PRICE-BASED APPROACH: A PROPOSAL

In a sense, the compliance mechanism of the protocol is at once both the cause of problems and the answer to those same problems. The quantity-based approach of Kyoto can be replaced with a price-based approach by the simple addition of a few lines of text to the compliance article. In particular, we propose allowing countries that have not sold their permits abroad to pay a fixed price per ton to a suitable international agent to cover any excess emissions at the end of the first commitment period in 2012.[3] We suggest a compliance payment of $50 per ton of carbon. The revenue generated from these payments would then be used to acquire emission reductions from other countries.

It is our belief that, in order to succeed, the Kyoto Protocol cannot demand strict emission limits regardless of costs. As noted above, in many countries, uncertainty about the cost of participation is one of the primary political/economic objections to the protocol. Compliance costs can be limited in several ways. The protocol already includes a number of flexibility mechanisms designed to reduce these costs; emissions trading, the CDM, joint implementation, inter-gas trading and the use of carbon sinks all help reduce the cost of compliance by increasing the range of options available to each country. Even if all of these features are effective, however, they only reduce costs, not limit them, nor do they reduce the uncertainty over them. Unexpected circumstances can still lead to high and possibly unacceptable costs associated with the Kyoto targets.

Our proposal specifically limits costs by providing a fixed-price alternative to domestic reductions or international permit purchases. Since countries can choose to pay the compliance payment on each ton of excess emissions, no country would be obliged to spend more per ton on GHG abatement than the established compliance payment. It is essential, however, to limit this option to countries that choose not to sell permits abroad in order to avoid excessive permit banking. The actual level of the payment must reflect a consensus among countries about how hard they are willing to work to reduce emissions. A high payment may not provide a useful compliance alternative. A low payment may lead some countries to put forth less domestic effort than the consensus desires.

A compliance payment is not the only way to limit costs. Another alternative is to require countries to subtract any excess tons at the end of 2012 from their commitment in the next budget period, perhaps with an additional tonnage penalty. Alternatively, countries could be required to solicit outside technical assistance in order to establish a more effective domestic policy. A third possibility is that a country with excess emissions could be required to demonstrate that it undertook appropriate policies and measures to try to meet its target. Mechanisms like these have the advantage that they do not necessarily require any extra-territorial payments by countries with excessive emissions. However, the mitigation incentives associated with these mechanisms are unclear and, depending on the actual design, could be too weak to encourage an adequate domestic effort or too onerous to limit costs.

A successful compliance mechanism should assure adequate domestic effort. The fundamental goal of the Kyoto Protocol is to reduce greenhouse gas emissions. With strict emission limits, it is relatively easy to verify whether a country achieves this goal: emissions are counted and compared to permit holdings. Alternative compliance options raise the question of how to determine whether a country has seriously tried to reduce emissions. In situations where obligations are not easily measured, participants could choose either to trust one another or to specify complex and invasive verification procedures.

Instead of requiring trust or invasive verification, our proposal provides economic incentives to ensure domestic effort. When properly implemented, these incentives encourage participating countries to meet the Kyoto goals based on their own self-interests. The advantage of the proposed approach is that the incentive is transparent and involves a minimum of subjective measures and administration.

The alternative of a fixed payment per ton of carbon creates incentives to identify and undertake emission reductions that are less costly than the payment. With a $50 compliance payment per ton, a country with 1 million excess tons could pay $50 million to remain in compliance. If additional domestic reductions are available for $20 per ton, however, the country could save $30 million by undertaking those domestic reductions and avoiding the compliance payments. While such incentives will not force a country to undertake every feasible reduction, they provide a constant pressure to do so. By using the revenues from the compliance payments to reduce emissions in other countries, we assure that participating countries either put forth the desired level of domestic effort or actually spend more to reduce emissions elsewhere.

To avoid any manipulation or abuse of the collected payments, a reverse auction would be held to purchase emission reductions in the second commitment period. Annex I countries as well as developing countries offering project-based credits approved through the CDM would be allowed to bid. The CDM will assure the legitimacy of the project-based credits in developing countries. Bids would consist of an offer to sell a specific number of Annex I permits or CDM emission credits at a specific per-ton price. This process is the reverse of a typical auction, in which participants offer to buy a product at a specific price; here, the participants offer to sell a product (emission reductions).

To complete the auction, an agent for the parties to the protocol would collect the bids and establish the successful auction price. The auction price would be a per-ton amount. Bids that offer permits at or below the auction price are accepted and receive the full auction price; the remaining bids are not accepted. The auction price would be chosen to exhaust the accumulated compliance payments while purchasing as many tons as possible. After establishing the price, the agent would direct funds from countries choosing to make compliance payments to the successful bidders. Because every bidder would receive the same price per ton, there is no incentive for a country to overstate its minimum acceptable price. A single bidder overstating its minimum acceptable price would not influence the eventual auction price but would reduce its own likelihood of winning the auction.

Importantly, this proposal would require a minimum of administration. Reporting requirements already established under the protocol track the permit holdings of Annex I countries. Existing mechanisms for evaluating CDM projects establish the legitimacy of developing country credits. The only administrative activities would be collecting the bids, establishing the consequent auction price and directing funds from countries making compliance payments to those with winning bids.

Our proposal would limit the cost of participating in the Kyoto Protocol, ensure that countries take suitable action and create clear incentives for GHG mitigation worldwide. The fixed compliance payment provides an unambiguous signal of the consequences of excess emissions. Countries could participate in the protocol and design domestic programs with complete certainty about their financial commitment and a clear incentive to reduce emissions. Because the financial incentive would apply to everyone, countries would be assured that all participants would make a reasonable effort to reduce emissions without any need for invasive or complicated domestic reviews. By using a reverse auction to distribute the revenue, the proposal avoids any opportunities to manipulate or misuse the revenue.

Despite the potential for relaxation of the Kyoto emission limits, adoption of the compliance payment mechanism would increase the likelihood of timely ratification and successful implementation. The proposed payment mechanism can accelerate the progress toward the longer-term goal of stabilizing atmospheric concentrations of GHGs.

4 POLICY SIMULATIONS

4.1 The Simulation Model

Moving to a price-based policy from a quantity-based approach gains certainty over cost at the expense of certainty over the number of tons to be emitted. Since each country's emissions are likely to exceed their allocated amounts if the cost of abatement on the world market exceeds $50 per ton, it is instructive to examine how much of the aggregate Kyoto Commitment would be met under the proposed price-based scheme.

To perform this analysis we have employed a simulation model based on the modeling results published in the special issue of the *Energy Journal*, May 1999 (hereafter *EJ*). Articles in this issue of the *EJ* report detailed costs of compliance estimates with the Kyoto Protocol for 12 different models[4] affiliated with the Energy Modeling Forum at Stanford University. In the introductory essay to the volume, Weyant and Hill (1999) state:

> the objectives of this study were the same as for previous EMF studies: (1) identifying policy-relevant insights and analyses that are robust across wide ranges of models, (2) providing explanations for differences in results from different models and (3) identifying high priority areas for future research. (pp. vii)

The simulation model employed in this chapter is based on the *EJ* results and is described briefly below; a detailed presentation can be found in Ghersi (2000).

Each of the modeling teams participating in the EMF effort was asked to analyse four distinct global GHG abatement regimes. These were: (1) a no trading regime where all countries were required to meet their target solely through domestic action, (2) an Annex B trading regime where countries of Annex B are permitted to trade allocated amounts, (3) the double-bubble which divides Annex B into two trading blocs – the EU and the rest of Annex B and (4) full global trading, which admits the full potential of the CDM to bring forth GHG reductions in the developing world. Each model analyses the four abatement regimes for each of four regions: the United States, the European Union, Japan and Canada, Australia and New Zealand – the CANZ group.

The results provided in Weyant and Hill (1999) permit one to extract pairs of marginal abatement cost and tonnes of GHG abated for each model, abatement regime and region. Ghersi (2000) extracts these pairs and uses them to calibrate the functions,

$$p_{ijk} = ax_{ijk}^{\ \ n}$$

where p_{ijk} are the marginal cost of abatement regime I in region j produced by model k, x_{ijk} are the GHGs abated with similar indexing and a and n are parameters. The calibrated coefficients are provided in the appendix.

Unfortunately, the four EMF regions leave out several Annex B countries, in particular, the former member countries of the Soviet Union – often termed the economies in transition (EIT) – and the rest of the world (ROW). To deal with the EIT, Ghersi (2000) constructs a linear marginal abatement cost curve from the EMF results. He uses the Annex B equilibrium trading price and the tons of GHG abated in the non-EIT Annex B countries to calculate the amount of GHGs that must have been abated in the EMF models by the EIT to hit the Kyoto target. This calculation yields a price–quantity pair through which one can pass a linear marginal abatement cost curve with a zero intercept. However, there is good reason to believe that some of the tons are 'hot air'[5] and that the marginal abatement cost of these tons is zero. Ghersi calculates the EIT abatement to be 474 million metric tons of carbon (MtC)[6] and assumes (with reasonable justification) that 200 million MtC of this is hot air.

As noted above, EMF results are supplied for a global trading regime, which generates a world equilibrium price. This price is passed to each marginal abatement cost curve for each region and the new curve developed for the EIT. The total of tons abated at that price in each region and the EIT is calculated and subtracted from the Kyoto target. The tons lacking are assumed to be supplied by the ROW through such things as the CDM. This calculation again provides a single point on a ROW marginal abatement cost curve and Ghersi passes a linear function through this point and the origin.

In summary, Ghersi has constructed for each of the EMF models, a marginal abatement cost curve for: the United States, the European Union, Japan, CANZ, the EIT and the ROW. For simplicity, and in an attempt to draw out some notion of central tendency, we average across all the individual model results in the simulations performed for this chapter.

There are a variety of reasons to believe that even with the best of market institutions, the EIT and the ROW will be unable to supply GHG reductions at the maximum amounts suggested by the EMF model runs. The likelihood that allocated amounts from the EIT and CDM tons from the ROW will be less than the maximum amounts is discussed by many of the authors of the *EJ* papers. To take this reality into account, we cut by half the amount of available EIT tons beyond the 200 MtC of hot air and the ROW tons were reduced by 75 per cent. These reductions were employed parametrically in all of our simulations.

4.2 The Simulation Results

We perform five simulations using the Ghersi model. The first simulation, referred to as the reference case, assumes that all four regions – the United States, the European Union, Japan and CANZ – meet their Kyoto target through a combination of domestic action, purchased credits (assigned amounts and CDM credits) and through purchased 'hot air'. In the other four simulations we model a compliance regime like the one proposed in this chapter, in which a country may be in compliance by meeting its target or by paying a fixed amount (the compliance payment – CP) for each ton emitted in excess of the Kyoto target. The four simulations differ by the amount of the charge ($25, $50, $75, $100).

Table 13.1 presents the results of the five simulations. In the Reference case, the four regions reach their Kyoto targets with a combination of 'hot air', imported credits and domestic action. The annual abatement cost for each region under each of the four alternative compliance payments is presented in the table.

Table 13.1 Annual compliance cost in millions of 1990 dollars

Region	$25	$50	$75	$100
European Union	5836	10486	14147	16520
United States	10037	17415	22909	26494
Japan	2215	4044	5588	6731
CANZ Group	1765	3075	4062	4670

Figure 13.1 presents the results of the five simulations in terms of greenhouse gas abatement. In the Reference case, the four regions reach their Kyoto targets with a combination of 'hot air' (22 per cent), imported credits (16 per cent) and domestic action (63 per cent). The marginal cost of abatement in this CDM-constrained global trading world is $122 per ton. A CP of $25 per ton causes the four regions to miss their targets by almost 40

per cent. However, at a CP of $50 per ton, almost 80 per cent of the commitment is reached and more than 90 per cent is reached at $100 per ton.

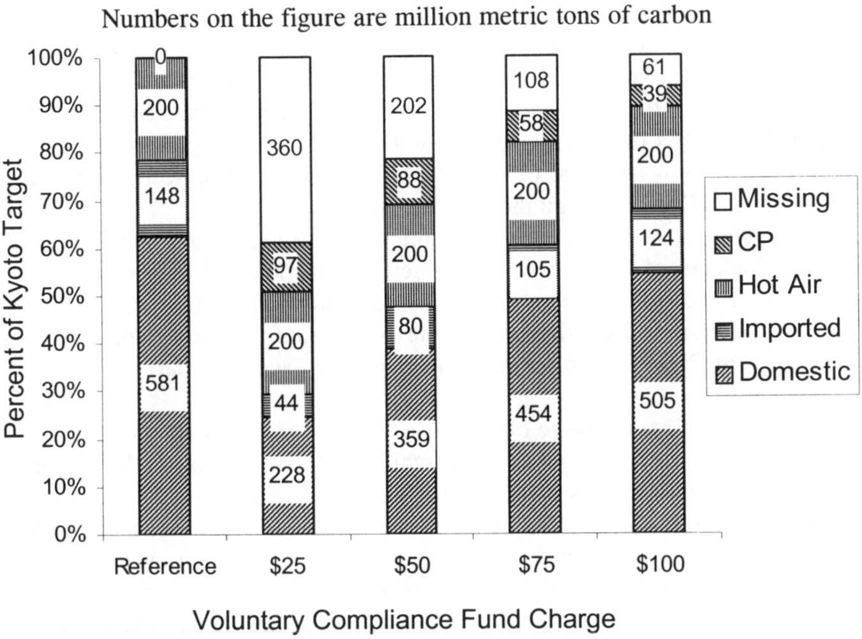

Figure 13.1 Greenhouse gas abatement in 2010

While even a $100 per ton CP charge does not guarantee that the four regions will meet their Kyoto commitments, it does guarantee a source of international finance for developing countries through the reverse auction discussed above. A CP of $50 per ton generates a fund of more than $17 billion to fund GHG abatement projects.[7] Over the long term, introducing inducements for developing countries to reduce GHG emissions will become ever more important.

4.3 Concluding Remarks

We have argued on both welfare theoretic and political grounds that the quantity-based approach of Kyoto is not only sub-optimal, but it in fact obstructs the very action that negotiators of Kyoto wanted – firm commitments to take action to limit GHG emissions. The fact is that the quantity targets leave the cost of Kyoto compliance unbounded and this makes it very difficult for many countries to agree to the commitment and ratify the protocol.

We have proposed a slight alteration to the protocol which serves to bound the cost of compliance with certainty and thereby remove what we believe to be the major barrier serving to thwart ratification. Our proposal adds a third leg to the compliance stool and allows countries to come into compliance with their Kyoto commitments by: (a) meeting their assigned amount target, (b) purchasing assigned amounts or CDM credits on the world market to meet the shortfall, or (c) paying a fixed compliance payment on all tons emitted in excess of the assigned amount (net of any tons purchased on the world market) to an agent of the UNFCCC to be used to secure assigned amounts or CDM credits on the global market.

The agent of the UNFCCC would solicit bids from Annex 1 countries and from providers of CDM credits in a reverse auction and then expend the funds collected from the compliance payments to purchase offered tons. The reverse auction is incentive-compatible with truthful bidding and ensures providers of CDM credits a market place without excessive market power on the buyer side.

Our proposed compliance option does not guarantee that all countries will hit their Kyoto targets. Of course, in reality, there is nothing in the existing compliance approach that makes such a guarantee. However, the simulation modeling reported in this chapter suggests that a compliance charge of $50 per MtC gets the four regions almost 80 per cent of the way toward meeting their Kyoto goals. Greater compliance charges would get closer to the Kyoto targets, while smaller chargers would lead to further departures from the targets. The 'right' level of compliance charge is ultimately a political decision.

APPENDIX

Table 13.A (a, n) *coefficients fitting the no trading and Annex B trading points in an* ax^n *functional form*

European Union

	a	*n*
ABARE-GTEM	0.4980	1.2770
AIM	0.3662	1.2100
CETA	n.a.	n.a.
G-Cubed	0.2668	1.1234
GRAPE	n.a.	n.a.
MERGE3	0.8143	1.0193
MIT-EPPA	0.0132	1.7300
MS-MRT	0.6902	1.0654
Oxford	0.0293	1.9069
POLES	0.1889	1.2799
RICE	0.2860	1.1697
SGM	0.0029	2.0177
WAGEM	0.0557	1.4356
Worldscan	0.3084	0.9990

United States of America

	a	*n*
ABARE-GTEM	0.3861	1.0922
AIM	0.2083	1.0963
CETA	0.307121111	1.015494199
G-Cubed	0.1672	0.9780
GRAPE	n.a.	n.a.
MERGE3	0.0749	1.3198
MIT-EPPA	0.0001	2.3928
MS-MRT	0.0448	1.3625
Oxford	0.0001	2.4731
POLES	0.0395	1.3302
RICE	0.1438	1.1570
SGM	0.0001	2.2773
WAGEM	0.0035	1.7714
Worldscan	0.0471	1.1252

Table 13.A *(Continued)*

Japan

	a	*n*
ABARE-GTEM	1.7430	1.3510
AIM	2.3177	1.0075
CETA	n.a.	n.a.
G-Cubed	0.0516	1.8005
GRAPE	0.050907829	1.589129617
MERGE3	1.5899	1.1471
MIT-EPPA	1.0549	1.3026
MS-MRT	0.0837	1.8734
Oxford	0.2547	1.8439
POLES	0.7108	1.3308
RICE	0.5122	1.3281
SGM	0.2368	1.4526
WAGEM	0.1908	1.6824
Worldscan	1.2879	1.0045

CANZ group

	a	*n*
ABARE-GTEM	3.0750	1.0904
AIM	2.1651	1.0029
CETA	n.a.	n.a.
G-Cubed	1.1114	1.0061
GRAPE	n.a.	n.a.
MERGE3	5.9271	1.0260
MIT-EPPA	0.0038	2.3301
MS-MRT	1.1411	1.2126
Oxford	n.a.	n.a.
POLES	0.5079	1.2844
RICE	0.5800	1.2249
SGM	0.2745	1.4270
WAGEM	0.0141	2.0767
Worldscan	2.1432	0.6156

NOTES

1. This chapter has been prepared for the 2000 Taipei Conference on Policies for Greenhouse Gases/Reduction and Pollution Control in Asian-Pacific Countries, 30 November–1 December, 2000, Institute of Economics, Academia Sinica, Nankang, Taipei, Taiwan. The chapter borrows heavily from the work of many groups and individuals. We are grateful for the ideas voiced by the Brazilian delegation to the UNFCCC; Warwick McKibbin and Peter Wilcoxen, Jennifer Morgan and Glenn Wiser, Marcus Schneider and countless others who have contributed to the debate on international climate policy. We also benefited from financial support generously provided by the Wallace Global Fund for a Sustainable Future and the German Marshall Fund of the United States. None of these groups or individuals is responsible for the contents of this document, however. This chapter reflects the opinion of the authors and does not represent the position of Resources for the Future, or the Centre International de Recherche sur l'Environnement et le Développement.
2. Article 3 states: 'The Parties included in Annex I shall, individually or jointly, ensure that their aggregate anthropogenic carbon dioxide equivalent emissions of the greenhouse gases listed in Annex A do not exceed their assigned amounts, calculated pursuant to their quantified emission limitation and reduction commitments inscribed in Annex B and in accordance with the provisions of this Article, with a view to reducing their overall emissions of such gases by at least 5 per cent below 1990 levels in the commitment period 2008 to 2012.'
3. This agent would be established by the protocol. It could be an entity similar to the CDM Board, or it might be a suitable private entity – for example, an international bank.
4. Two of those models, GRAPE and CETA, giving estimates for one zone only (Japan and the United States respectively), were dropped. POLES, a French model not part of the EMF, is added to the study based on data provided by its modelers.
5. A term used to suggest that these are not real reductions or similarly, the marginal cost of reducing a hot air ton is zero.
6. This is the average calculated across all the models.
7. Our simulations suggest that this fund grows to $18 billion when the VCP is $75 per ton, but declines back to the $17 billion range at $100 per ton.

REFERENCES

Ghersi, Frédéric (2000), 'Supplementarity and Compliance to the Kyoto Protocol: Assessment of Policy Options', Paper presented at the RFF-CIRED workshop part II, 1 September, Paris.

Pizer, William A. (1996), 'Modeling Long-Term Policy Under Uncertainty', PhD, Harvard University.

Pizer, William A. (1998), 'Prices vs. Quantities Revisited: The Case of Climate Change', Discussion Paper 98-02, Washington, DC: Resources for the Future.

Pizer, William A. (1999), 'The Optimal Choice of Climate Change Policy in the Presence of Uncertainty', *Resource and Energy Economics*, **21** (3–4), pp. 255–87.

Pizer, William A. (2001), 'Comparing Price and Quantity Controls for Greenhouse Gases', in M.A. Toman (ed.), *Climate Change Economics and Policy*, Washington, DC: Resources for the Future.

Pizer, William A. (2002), 'Combining Price and Quantity Controls to Mitigate Global Climate Change', *Journal of Public Economics*, **85**, pp. 409–34.

Weitzman, Martin L. (1974), 'Prices vs. Quantities', *Review of Economic Studies*, **41** (4), pp. 477–91.

Weyant, John P. and Jennifer Hill (1999), 'The Costs of the Kyoto Protocol: A Multi-Model Evaluation, Introduction and Overview', *The Energy Journal*, Special Issue.

14. A better alternative to the Kyoto Protocol[1]

Warwick J. McKibbin

1 INTRODUCTION

The election of the Bush Administration in the United States has apparently transformed the climate change debate. In an announcement that generated worldwide condemnation, President Bush declared early in 2001 that the United States would not support the Kyoto Protocol. Commentators seemed surprised by this development although it was inevitable that at some stage a key player in the climate change negotiations was going to question the approach of the Kyoto Protocol. After all, by May 2001 it had been more than three years since the negotiation of the Kyoto Protocol at COP3, which was held in Kyoto in December 1997. Intense negotiations over the intervening period have focused on how to implement the Kyoto Protocol and these seemed to have stalled at COP6 in The Hague in November 2000. The Kyoto Protocol has been signed by 84 countries but not yet ratified by enough of the key countries and the United States has apparently rejected it.

Why has it been so difficult to take the next step of implementation? The simple answer is that design of mechanisms within the Kyoto Protocol has generated too much complexity and requires too many new international institutional developments to be plausible. The fundamental answer is that the Kyoto Protocol was never going to work because it is the wrong approach to tackling the climate change issue. President Bush has now created an environment for designing a more sensible approach to climate change policy. The approach presented in this chapter is one of the main alternatives to the Kyoto Protocol.

In going back to consider alternatives to the Kyoto Protocol, it is worth focusing on what climate change policy is fundamentally to address. The core issue about climate change is how to design a policy response in an environment of considerable uncertainty. There is enough evidence and professional expertise to suggest that climate change could be a serious problem. What is required is an insurance policy against the *possibility* that climate change could be very costly to the planet. The key issue is how much insurance should humanity pay, given the current state of our understanding.

The answer is that we do not really know what cost we should pay now. Neither do we know by how much nations should reduce carbon dioxide emissions or how quickly.

Key elements of the McKibbin – Wilcoxen Proposal

All countries create two assets:
- an *emission permit* which is required by fossil fuel industries to supply a unit of carbon annually;
- an *emission endowment* which gives the owner an emission permit every year forever.

All countries create two domestic markets:
- a domestic *emission permit* trading system with a fixed price of $US10 per ton of carbon in Annex I countries and a cap price of $US10 in non-Annex I countries;
- a domestic *emission endowment* trading system with a flexible price.

In 2000, all countries are allowed to make a one-off allocation of emission endowments domestically based on Kyoto targets for Annex I countries and current emissions plus x% for non-Annex I countries. Trading in both markets begins in January 2001.

Permits must be reconciled against production or imports of carbon on an annual basis at the top of the carbon production chain – coal mines, oil refineries, gas refiners. Production that is exported is exempted.

Every decade there is a meeting of the Conference of the Parties to the UNFCCC to evaluate the extent of abatement and the climate science and to negotiate a new price for permits.

Nonetheless, the Kyoto Protocol consists of a specific set of targeted reductions in emissions: 5.2 per cent for Annex B countries (essentially OECD economies plus countries of the former Soviet Union) relative to current emissions within the next decade (or more precisely, between 2008 and 2012). The target was set although the negotiators had no way of knowing how costly it will be to attain it. Understandably, therefore, countries are reluctant to implement a policy that could potentially be very costly and where the benefits are also uncertain. Although there is some important flexibility built into the Protocol to smooth costs across countries, the total cost results from the overall targets. Very few countries would actually implement a policy with such uncertain costs, once the critical time of implementation was reached.

Just as important as the uncertainty of costs, only a subset of countries are part of the agreement and those countries are expected to create new international institutions and laws that can accommodate the various mechanisms at the foundations of the Protocol. The most problematic are international trading of emission permits, which requires a system of monitoring and enforcement that is unlikely to be feasible in the near future, and the Clean Development Mechanism, which requires detailed and costly evaluation of carbon-reducing investment proposals in developing countries on a project-by-project basis.

So what can be done? A number of realistic proposals have been made. One from the Resources for the Future (see Kopp et al., 1997, 1999) would place a cap on the prices of emission permits that each nation would issue. This would guarantee that the implementation cost of the Protocol would not exceed a given level. An alternative is the McKibbin–Wilcoxen (MW) Proposal from the Brookings Institution (1997a, 1997b, 1999, 2002), which proposes a a fundamental re-thinking the approach of fixed targets and the international trading of emission permits that is embodied in the Kyoto Protocol.[2] Both proposals have evolved over time and can be considered as 'early action policies', while countries still attempt to solve the problems with the Kyoto Protocol. This chapter lays out the key features and advantages of the Blueprint Proposal and its attractiveness as an early action policy. It then compares this proposal to the approach of Kopp et al. (1999) which, although apparently a similar idea, has a few fundamentally different implications.

2 WHAT IS WRONG WITH THE KYOTO PROTOCOL?

The objective of the Kyoto Protocol is to impose binding greenhouse gas (GHG) emission targets for the world's industrial economies and former communist economies of Europe ('Annex I' countries) to be achieved by the period 2008–2012. By directly binding emissions, policy makers presumably believed that they could achieve the goals of the UNFCCC through political commitment. Clearly, this was perceived to be the easiest approach to follow because explicit targets can be negotiated and monitored. Given that fixed targets for emissions by Annex I countries have been agreed, although not yet ratified in key countries,[3] the main issues currently being debated are how to minimize the costs of the Kyoto Protocol and how to bring developing countries into the agreement.

The issues of cost minimization and developing country participation are clearly recognized in the Kyoto Protocol. Costs are addressed through provision for international trading of emission allowances among the countries that accept binding targets. In addition, the Protocol provides for a 'Clean Development' Mechanism, under which agents from industrial countries can earn emission credits for certified reductions from investments in 'clean development' projects in developing countries that have not taken on binding targets.

The first problem with the Kyoto Protocol is the focus on achieving rigid targets and timetables for emissions reductions at any cost, rather than substantial reductions in emissions at reasonable cost. Some negotiators at Kyoto understood the problem with fixed targets and thus flexibility mechanisms, such as permit trading, were included in the Protocol. A crucial but mostly ignored issue is that any fixed targets, for the world or for a group of countries, even differentiated targets, are likely to be inefficient because we really don't know what these will cost over the long period of time being discussed.[4] If the actual costs of abatement turn out to be much larger than estimated it is unlikely that countries will continue to adhere voluntarily to the Kyoto Protocol. Some form of extreme enforcement mechanism needs to be designed to hold the Protocol together. Imposing arbitrary but binding targets on developing countries is even more problematic because there is even greater uncertainty about what the appropriate targets should be. An overly tight target will cause countries to depart from the agreement and an overly loose target will mean that low-cost opportunities will have been lost.

Permit trading within the Kyoto Protocol is essential to minimize (but not eliminate) these problems. However even a permit trading system could be problematic. In a series of papers (McKibbin and Wilcoxen, 1997a, 1997b) we have pointed out that under some plausible scenarios for the future evolution of the global economy, the economic pressures caused by the large transfers of wealth internationally that underlie the claims over permits, could cause severe fluctuations in real exchange rates and international capital and trade flows. Whether or not this actually emerges as a future problem will depend on a number of factors, especially the ultimate price of permits and the initial allocation of permits. In particular, this may be a problem if permit allocations are used excessively as a way of persuading countries to participate in an agreement. Although there is uncertainty about whether this effect is large or small, the main point is that we can't be sure that the economic problems we highlight will not emerge in the future.

Another problem with permit trading under the Kyoto Protocol is that the price of permits for all countries depends on the demand and supply of permits by all countries. If one participating country cheats, then the value of permits for all countries will be affected. If a large country cheats, then the value of permits will be debased and the system is likely to collapse. There is currently no international rule of law that can prevent this from happening nor is it easy to see what credible penalties could be imposed to prevent this from happening under all possible scenarios. It is also hard to imagine why developing countries would want to participate in a centralized system like the Kyoto Protocol especially once the enforcement mechanisms are made explicit and without knowing the possible costs of accepting a binding emissions target.

Overall, it seems that both politically and economically there may be potential problems with the Kyoto Protocol involving the possibly large wealth transfers between economies. More fundamentally, the incentives of key players are not clearly consistent with the Protocol under extreme developments, without some, yet to be identified, enforcement (and monitoring), mechanism.

3 BASIC FEATURES OF THE McKIBBIN–WILCOXEN BLUEPRINT PROPOSAL

Rather than centralize the process of reducing carbon emissions and creating new international institutions, it is better to coordinate responses across countries (what Richard Cooper of Harvard calls an approach of agreed actions) in an explicit way so that each country would pay the same price for emitting carbon. Furthermore, it is appropriate at this stage to create property rights over emissions of carbon dioxide from burning fossil fuels only. As much as it would be nice to have alternative gases and sinks, as in the Kyoto Protocol, it is an administrative nightmare to deal with them in the near term and adds enormous complexity to the task.[5] In the future, these could probably be added to our system on a country-by-country basis without compromising the system.

There are a number of goals that should be at the core of any climate change regime. The first is to put in place long-run targets for emissions. The second is to guarantee the short-run costs for industry and consumers. The third is to create domestic institutions that allow people to self-insure against the uncertainties created by climate change. There should be market mechanisms that give clear signals about the current and expected future costs of carbon. There should be coalitions created within countries with the self interest of keeping climate change policy from collapsing rather than creating a system of international sanctions in order to sustain the system.

The key innovation of the Blueprint Proposal is that in order to achieve the objectives of long-run targets as well as guaranteed short-run costs we would create two emissions-related assets and associated markets for both in each country. These markets are not linked internationally and therefore there is no international trading of these assets. The two assets are designed to set a long-run goal for emissions and to limit the short-run costs. Fortunately the two markets also create a mechanism for managing risks associated with climate change policy within each economy so that very little else needs to be done to implement a consistent and simple market-based approach to tackling the climate change issue.

The first asset is an *emission permit*. This certificate would entitle its holder to produce a unit of carbon each year (each permit would have a date stamp and be valid only in the year issued). These emission permits are to be held by producers of carbon to acquit for each unit of carbon produced each year. The holder of the permit and the ultimate owners who supply the market need not be the same people.

The second asset is an *emission endowment* (or *perpetual permits*), which is a certificate that entitles the holder to an emission permit *every year forever*. The emission endowment is like a government bond that pays a coupon of an emission permit every year. Another way to think about the two assets is that the emission endowment is like stock in a corporation whereas the emission permit is the dividend the corporation pays each year to people who hold the shares. The stock value is the expected value of future dividends.

There is a critical difference between the two asset markets. The endowment market, which is a long-term market, would be one in which the supply of carbon is fixed (the goal of policy) but the price is flexible. The government cannot issue more endowments after the initial allocation but can buy back endowments in future years if the target for emissions is to be tightened. Because the endowment is the perpetual lived asset, its price will reflect the expected future price of emission permits in each year (analogously to the stock price and the dividends of a company).

We treat the market for emission quite differently. The demand for permits will be by industry that requires a permit for producing carbon. The supply of permits in any year will be composed of the permits generated by the emission endowments. We also introduce a supply of permits by the government of a sufficient amount to hold the price of permits at an internationally agreed level. Thus the price of permits is fixed and the number of permits (equal to the amount of carbon emissions in any year) is allowed to vary. Every ten years there would be a negotiation between all countries in which the price for emission permits is agreed to and fixed for ten years. The price of permits would be fixed in each economy by governments selling additional permits into the market after the permits generated by the endowments have been fully utilized. Thus a producer that wants to produce a unit of carbon for domestic use can get a permit in a given year by either having an existing emission endowment, purchasing an emission endowment in the endowment market (this would be sold by another private holder of an endowment), or purchasing an emission permit in the permit market which is either supplied by a private owner of a permit or from the government.

We propose that the initial price of the annual permits, which would determine the marginal cost of emitting carbon, be set at $US10 per ton of carbon (in 1990 dollars). The price would be the same in all markets in all participating countries and thus the cost of removing carbon at the margin in each economy would be identical in the short run. No complicated system of international trading in permits or global monitoring would be required –

addressing a central flaw in the current Kyoto Protocol. Moreover, the value of permits in the United States or Australia or China would not depend on how permits are generated in other countries.

In contrast, the price of endowments would be flexible, reflecting the outcome of market forces, the period of fixed permit prices in the near future, as well as the expectations of private actors as to what is likely to happen after the current negotiation period. Industry and consumers would be expected to respond to both the short-run price signals (which are known for ten-year periods) as well as the long-run price signals (which are market determined) in making spending and investment decisions.

The purpose of separating the endowment market from the emissions market is to ensure that, over the long run, emissions do not exceed a given limit. The annual emissions permitting process cannot accomplish this objective since it operates on the basis of a fixed price (the emissions fee), not a fixed quantity.

The initial allocation of endowments would be up to each government. We propose giving a significant portion to every person in an economy at a point in time. We would also give a large amount to fossil fuel industries as compensation to shareholders for the capital losses of significant structural change as well as to workers in these industries. The initial allocation of endowments will create a natural constituency supporting climate change policies because the value of the endowments in future years will depend on the commitment of the government to pursue sound environmental policies. This would create a mechanism for enforcement of the agreement that is internal to each country.

4 HOW CAN DEVELOPING COUNTRIES BE INDUCED TO PARTICIPATE?

It is important to distinguish between Annex B (industrial) countries and developing countries. Failure to do so would unduly inhibit the growth of the developing world and would not attract their support for a global system that is absolutely crucial for a successful policy.

Accordingly, it is appropriate in the case of Annex B countries to use the Kyoto targets as the endowment allocation within each economy. For developing countries, however, it is only reasonable to allow endowments far in excess of current requirements (the precise levels being subject to international negotiation). With endowments greater than requirements for permits over the next several decades, the price of permits in these countries would be zero and thus there would also be no short-run costs. In contrast, the price of endowments in developing economies would be positive, since the price would reflect the expected future price of permits. Thus, a price signal can be introduced to the developing world that will affect current investment plans without entailing short-run costs.

A developing country can therefore begin to contribute to a reduction in emissions with a firm commitment in the form of endowments. This reduction will be realized, however, only when emissions actually bump up against the endowment limit. The faster a country's economy grows and thus the faster pace at which emissions are growing, the more rapidly the endowment constraint will become binding.

Meanwhile, carbon-intensive industry will have fewer incentives to shift from Annex B countries into developing countries in order to avoid the carbon charge in industrial countries because they would need to consider the fact that all countries will be participating in the overall emissions reduction program (of which endowments would play an important role). The differential endowment system – one for industrialized countries, another for developing countries – would also have the added benefit of factoring in the cost of emissions in decisions by foreign private investors when deciding whether to commit funds to developing countries. Another important issue that our approach avoids relative to Kyoto is that under Kyoto if a developing country joins the Annex B trading system they will immediately face the new higher price of carbon, which will lead to immediate structural change within that economy. Even though compensation could be made through giving a large allocation of permits, it is not clear that the officials who receive the permits will be the same people who are impacted by the jump in carbon prices.

Overall, the nationally based emissions permit and endowment program is far more appealing than the Kyoto Protocol. All institutions are created and managed within each economy. Breakdowns in the infrastructure of any given market will not spill over to markets in other countries. To be sure, there would be fluctuations in the amount of global emissions, but the critical result is that the variations would be around a downward trend. Furthermore, decentralizing responsibility for taking action to individual countries would make the whole program more sustainable than the Kyoto alternative, which requires participation by all countries in an international permit-trading regime.

Another advantage of the approach proposed here is that the negotiation every decade on the permit price allows a great deal of flexibility. Monitoring of emissions and the extent of induced abatement activities can be undertaken more easily than in a global program. If information changes then the price of permits can be changed by international agreement. The endowment market will reflect this information immediately and will enable more rapid but cost-minimizing adjustment if required.

5 AS AN EARLY ACTION PROPOSAL

Finally, the permit and endowment approach can and should be easily implemented in the United States and all other countries as an early action policy. By establishing such a system with a low initial price for permits, all domestic institutions that would be required if the Kyoto Protocol proceeds, would be created in the meantime. To move from the fixed price system that we propose to a flexible price system under the Kyoto Protocol, all that is required is to remove the government intervention from the permit market in 2008 and allow international trading of the permits then. Alternatively and more likely, countries that implement the Blueprint Proposal would find that it works so well in providing price signals to consumers and industry that there will be no need to move to the Kyoto-style system over coming years.

The idea of a fixed price system appears similar to the RFF proposal of Kopp et al. (1999). However, there are some important differences. A key difference is that we design a system with a very long horizon, which creates clear property rights over carbon emission well into the future, whereas the RFF proposal is a short-term cap within the constraints of the Kyoto Protocol. There is nothing to determine what happens after the first budget period of 2008 to 2012. Secondly, we would create a set of domestic institutions which could be done today based on existing rules of law in each country. The RFF proposal requires the creation of international institutions just as the Kyoto Protocol requires, which is why it is stalled. Once created, these international institutions would essentially become redundant once the price cap was reached since international trading would not be required once the cap was reached. However, it is unlikely that, once created, these international institutions would not want to play an integral role in country systems. Indeed there is a great deal more uncertainty under the RFF proposal because of the short-term nature of any property rights created and the fundamental international integration that our system seeks explicitly to avoid. If a large country pulls out of the RFF system, it is more likely to collapse because of the potential tradability of the international permits.

Therefore, there are a number of similarities between our approach and the RFF approach in a world of perfect certainty, primarily the idea of a fixed price target rather than a fixed target for the quantity of emissions. However, once uncertainty – the fundamental issue behind climate change – is introduced, the two approaches are different in important ways. We would essentially jettison much of the Kyoto Protocol (apart from the Annex B targets as a basis for the creation of endowments, although these could go as well if it were politically possible to renegotiate these targets) whereas the RFF proposal would function as a safety valve within the Kyoto Protocol framework.

6 CONCLUSION

The key objective for those interested in promoting responsible climate change policy is to allow each country to administer its program, without depending on other countries, but within an overall cooperative framework that provides constructive incentives for private actors to control emissions in an efficient manner. The McKibbin–Wilcoxen proposal outlined here would accomplish this objective, ensuring sufficient flexibility for private actors, providing incentives for developing countries to commit to the system and creating constituencies within all countries to sustain the agreement, all without the need for cross-border intervention.

Finally, raising the price of carbon by a known amount in the short run would establish an insurance premium to be paid for climate change prevention over the coming year, while reducing the short-run uncertainty for investment planning, and creating a market that accurately prices carbon emissions for long-run planning purposes. Credible price signals can guarantee that emissions of carbon will be lower than otherwise would have been the case. Perhaps emissions will not be low enough as time proceeds and as we obtain better information and improved climate science, but a flexible system of emissions reduction can deal with this over time. Starting with small, but significant action now is far better than continuing to argue over the Kyoto Protocol and failing to implement policies that could make a meaningful start toward emissions reduction. The current situation with climate change policy based on the Kyoto Protocol generates enormous uncertainty for investment decisions and compounds the cost of climate change.

ACKNOWLEDGEMENTS

The author thanks the editors for helpful suggestions and the Brookings Institution for financial support. This chapter draws on joint research with Professor Peter Wilcoxen of the University of Texas at Austin. The views expressed in this chapter are those of the author and should not be interpreted as reflecting the views of any funding organizations or the organizations with which he is affiliated.

NOTES

1. Revised version of a chapter prepared for the 2000 Taipei Conference on Policies for Greenhouse Gases Reduction and Pollution Control in Asian-Pacific, at the Institute of Economics, Academia Sinica, 30 November–1 December, 2000.
2. The Blueprint Proposal, in some ways, is similar to the global carbon tax proposal of

Professor Uzawa (2000) with the crucial difference being that most of the revenue generated by raising the price of carbon in the Blueprint Proposal does not go to the government but is kept within the private sector.

3. As of September, 84 countries have signed but only 29 of the low emitter countries have ratified.
4. See McKibbin and Wilcoxen (1997a) and Kopp et al. (1997) for arguments about the difference between price and quantity caps under uncertainty.
5. A cynic might think that, although a good theoretical case can be made for including multiple gases and sinks in a comprehensive policy, they were actually included to kill the Kyoto Protocol because of the impracticalities of including them both through the imprecision of current techniques of monitoring them or the high transaction costs of doing so.

REFERENCES

Kopp, R., R. Morgenstern and W. Pizer (1997), 'Something for Everyone: A Climate Policy that Both Environmentalists and Industry Can Live With', 29 September, Washington, DC: Resources for the Future.

Kopp, R., R. Morgenstern, W. Pizer and M. Toman (1999), 'A Proposal for Credible Early Action in US Climate Policy', Resources for the Future, Weathervane, available at www.weathervane.rff.org\features\feature060.html

McKibbin, W.J. and P. Wilcoxen (1997a), 'A Better Way to Slow Global Climate Change', *Brookings Policy Brief*, No. 17, June, The Brookings Institution, Washington, DC, [online] available at http:\\www.brook.edu\comm\PolicyBriefs \pb017\pb17.htm

McKibbin, W.J. and P. Wilcoxen (1997b), 'Salvaging the Kyoto Climate Change Negotiations', *Brookings Policy Brief*, No. 27, November 1997, The Brookings Institution, Washington, DC, [online] available at http:\\www.brook.edu\comm\ PolicyBriefs\pb027\pb27.htm

McKibbin, W.J. and P. Wilcoxen (1999), 'Designing a Realistic Climate Change Policy that includes Developing Countries', paper prepared for the United Nations University Symposium on Global Environment and Economic Theory, 24–25 October, in Tokyo, [online] available at http:\\www.msgpl.com.au\msgpl\ download\developing.pdf

McKibbin W.J. and P. Wilcoxen (2002), *Climate Change Policy after Kyoto: A Blueprint for a Realistic Approach*, Brookings Institution Press.

Uzawa, H. (2000), 'Global Warming and Intergenerational Equity', paper presented at the United Nations University Workshop on Global Environment and International Cooperation, 23 March, in Tokyo.

15. Joint Implementation, the Clean Development Mechanism and the baseline: an economic analysis

Ekko C. van Ierland, Rianne de Leeuw and Joram Krozer

1 INTRODUCTION

The 1997 Kyoto Protocol established the flexible instruments Joint Implementation (JI) and the Clean Development Mechanism (CDM) to combat climate change. These project-based mechanisms can enhance the cost effectiveness of greenhouse gas (GHG) reduction because of differences in emission reduction costs in various countries in the world. For both joint implementation and the clean development mechanism it is essential to analyse to what extent a project will lead to additional greenhouse gas emission reduction, as compared to the baseline.

The advantages of JI and CDM are widely recognized, but serious problems may occur on their implementation. Clear methods should be developed to establish the net additional emissions reductions for various types of projects under JI or CDM. Particular attention should be given to methods for establishing the baseline and to the possibilities of carbon leakage, that is the option that additional GHG emissions will occur elsewhere, as an undesired side effect of a project.

Although some developing countries show a reluctance to participate in CDM projects, it seems that more and more are getting interested because of the opportunities for additional financing of projects that foster economic growth and contribute to sustainable development. Most developing countries, however, would like to avoid that if they will eventually have to fulfill their own GHG emission reduction targets, the 'low hanging fruits' have already been picked and only the very expensive policy options remain.

Various authors have discussed the issues or problems surrounding JI and CDM. Woerdman (1997) notes that conceptual and practical problems, like baseline setting, might hinder the implementation of international instruments and the cost effectiveness of these instruments. JIRC (2000) uses the results of the re-estimation/verification of ten Dutch AIJ[1] projects to address problems of baselines and verification and to make recommendations for improvement. One recommendation is to use project-specific baselines and to develop more precise guidelines for baseline setting. Other studies on this topic (often based on AIJ projects) are Jepma et al. (1998), OECD (1999), Ellis and Bosi (1999) and Ellis (1999).

This chapter discusses some important methods for defining the baseline and provides the results of four case studies for different options for reducing emissions of greenhouse gases. The aim of the chapter is to compare the methods of baseline setting for a variety of emission reduction options, including renewable energy, fuel switch, waste management and agro-forestry. We compare top-down approaches with bottom-up results and discuss some of the complications of calculating baselines on a project basis. In section 2 we describe top-down and bottom-up methods for defining the baseline. Sections 3–6 cover the case studies for, respectively, the Philippines, Egypt, Ecuador and Mexico. Section 7, finally, contains the discussion and conclusions.

2 METHODS FOR ESTABLISHING BASELINES

The definition of the baseline is paramount in calculating the emission reductions of JI or CDM projects (Michaelowa, 1998). The baseline is important in the allocation of emission reduction certificates and in determining the allocative efficiency of JI and CDM. This section will explain the issue of baseline setting and will present a model that can quickly indicate the emissions reductions associated with a project. The model is used to analyse the effect of different baseline approaches on the calculated greenhouse gas emission reductions.

The baseline for a JI or CDM project represents the estimated future emissions at the project location in absence of the investment under the project (Van der Gaast and Woerdman, 1999). It is an important concept, since it is used to calculate the emission reductions achieved in a project, by comparing the emissions in the baseline with the emissions in the project. In this way the baseline determines the allocation of emission reduction certificates. Many analysts regard the development of methodologies for setting baselines as the most difficult task in the specification of rules for CDM (Stewart, 1999). This is mainly because of the hypothetical character of the baseline situation.

Over the past few years, a range of baseline approaches has emerged in the literature. Figure 15.1 gives an overview of some of the possibilities in baseline setting as found in literature. The figure shows project-specific, hybrid and multi-project baseline methods as main categories. A *project-specific* baseline is determined on a case-by-case basis, with project-specific measurements or assumptions for all key parameters. Each baseline is used only for the project for which it was developed (Bandsma et al., 2000). In a *hybrid* baseline, one or more key components of the baseline are standardized. It can be applied to a (limited) number of projects. Hybrid baselines have a greater degree of standardization and aggregation as project-specific baselines, but are less standardized than multi-project baselines.

According to Ellis and Bosi (1999), the local and regional variability of the different baseline components will determine the extent to which standardization is feasible. A *multi-project* baseline consists of standardized and aggregated key assumptions. It can be applied to a number of similar projects, for instance within the same sector of a country or region. It sets a standard baseline value for a specified technology or industrial sector, at a selected level of aggregation (Bandsma et al., 2000).

In literature, baseline categories are not yet defined in a standard manner. The most notable difference is the treatment of top-down and multi-project approaches. Often, the top-down approach is seen as a multi-project baseline (for example Ellis and Bosi, 1999; OECD, 1999). In other instances, the approach is treated as a separate category (Bandsma et al., 2000). Generally, top-down baselines are highly aggregated. They could be used to assess emission reductions resulting from policy initiatives or national commitments, but not for very specific project activities. Baselines that are derived in a top-down manner are not necessarily projections of the future. Once the aggregate baseline is set, baselines for individual projects (or types) can be allocated on the basis of the aggregated baseline. In Figure 15.1, top-down baselines are treated as special types of multi-project baselines.

It is important to recognize that even under each separate approach there is a wide range of opportunities to establish a baseline. For example under the benchmark approach, one can choose to set the benchmark at a specific level, such as the best available technology, country average, OECD average or recent investments. For each baseline, the reference used can be historically based or based on a future (projected) situation. Finally, a baseline can be set *ex ante* (before the start of the project) or *ex post* (after implementation). These choices will significantly affect the amount of emission reduction that is attributed to a project.

To simplify the numerous baseline approaches, a model was developed (de Leeuw, 2000) to get a first (reliable) indication of the baselines and the emission reductions that can be expected in JI or CDM projects. It is used as a framework to analyse the effect of different baseline approaches. The starting point of the model is that the emission reductions of a project are calculated by the difference between the project emissions and the baseline emissions.

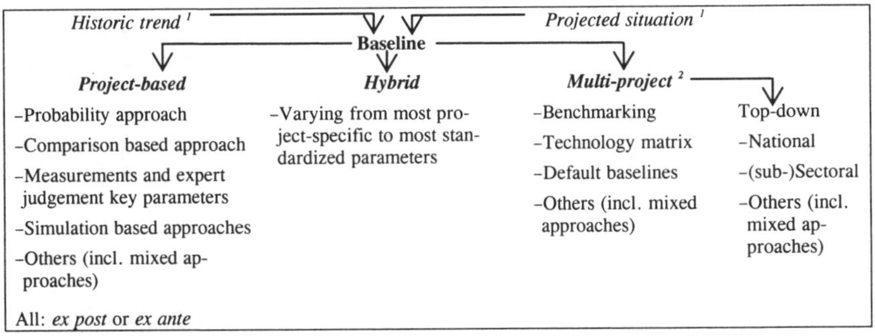

Notes:

[1] Authors' addition.

[2] In literature, the multi-project and top-down approach are defined in varying ways.

Sources: Based on OECD (1999), Ellis and Bosi (1999) and Meyers (1999).

Figure 15.1 Possibilities in baseline setting

The model identifies two aspects in establishing the *baseline emissions* of a project, that is the *data level* and the *reference used*. First, the data level can range from project-specific (bottom-up) to top-down. A top-down approach uses national data as a basis to calculate emissions at the project site. A bottom-up approach uses project-specific information. Second, the reference for the baseline can range from a historic trend to a projected future situation. Using historic trends, one calculates the baseline using historic trends (extended into the future) for the variables. It is assumed that these trends are a good indication of what would happen in the future if the project were not implemented. When a future situation is used as a reference, one defines a situation that is expected and/or strived for and, accordingly, calculates the baseline emissions. This allows for a baseline that anticipates changes in trends, which is very useful, for example, in rapidly developing countries.

Combining the two variables, the model specifies four possibilities to calculate a baseline for a CDM project: (1) bottom-up based on historic trends; (2) bottom-up based on a future situation; (3) top-down based on historic trends; (4) top-down based on a future situation. These methods comprise the various alternative methods found in the literature.

The second determinant of a project's emission reductions is the emissions associated with the project itself. The model specifies two ways to calculate these *project emissions*, namely top-down approaches (using national data) or bottom-up approaches (using project data). In practice, a top-down approach can only be used for projects with a very large scale (for example national electricity generation changes) or for national policy actions. In most cases, a top-down approach for assessing project emissions will be inappropriate. Summarizing: the model defines four ways to calculate baseline emissions and two ways to calculate project emissions.

Table 15.1 Model to categorize baselines and emission reductions for a JI or CDM project in various ways

		Baseline emissions			
		Historic trend		Future projected situation	
		Top-down	Bottom-up	Top-down	Bottom-up
Project emissions	Bottom-up	Emission reduction (1)	Emission reduction (2)	Emission reduction (3)	Emission reduction (4)
	Top-down	Emission reduction (5)	Emission reduction (6)	Emission reduction (7)	Emission reduction (8)

Note: Emission reduction = baseline emissions − project emissions.

To calculate the emission *reductions* of a project, one has to subtract the project emissions from the baseline emissions. This gives eight ways to calculate the emission reductions of a project. The numbers 1–8 in Table 15.1 represent the eight possible ways to calculate the emission reductions of a project. Emission reductions are calculated by subtracting the project emissions from the baseline emissions, both calculated according to the scenario chosen. For case number (1), the baseline emissions are calculated using historic trends and national data (top-down). The project's emissions are calculated using project-specific data. The resulting, calculated emission reductions for case (1) are established by subtracting these two emissions estimates. The same reasoning applies to the other seven cases specified in the model.

A JI or CDM project may alter the economic development in the project area. Since the model decouples the situation *with* and *without* the project, it can handle these so-called rebound effects.

Investors as well as verifying companies can use the model to review the emission reductions associated with a project. The information required for the model is modest and part of it is publicly available (for example on the internet). This enlarges the options for third-party review of the baseline and claimed reductions. For example, to establish a top-down baseline for energy projects, national data is needed on population, production and emissions (from fossil fuel use or process emissions). For other project types, industry data are needed on production and emissions. However, to calculate bottom-up baselines and the bottom-up project emissions, specific project data is necessary, for instance projected emissions (and/or technical specifications), population in the project area and, preferably, GDP in the project area. To acquire these data, there is a strong need to rely on studies and data provided by the investor. In both the top-down and bottom-up approach, the issue of lifetime of the project and its capital goods has to be solved, as well as the issue of leakage.

There are many more ways to calculate emission reductions besides those specified in the model. Even under each of the eight cases of the model, an immense range of alternatives exists. Accordingly, the goal of the model is not to cover the whole area of baselines, but to provide a structured way to calculate emission reductions from varying perspectives, giving a first indication of the amount and likely range of emission reductions achieved through the project. This is useful to both investors and verifying agencies.

The next sections discuss four case studies in which baselines were established and emission reductions calculated. The main questions underlying these case studies are how emission reductions can be calculated, how the baseline should be established and which complications occur in this respect for practical implementation of CDM and JI.

3 PHILIPPINES CASE STUDY: RENEWABLE ENERGY

This case study involves the electrification of a village in the Philippines that is not connected to the grid. The electrification will be realized using renewable energy sources. The project involves a pilot, possibly paving the way for more extensive electrification. The project includes two phases. In the first phase, a PV/LPG hybrid (Photo Voltaic combined with a diesel generator for back-up) is installed in combination with a mini-grid. In the second phase, the hybrid is replaced by a biomass generator, which will be plugged into the grid. The total lifetime of the project is 20 years; one year for the first phase and 19 years for phase two.

In short, the baseline comprises two parts: (i) the emissions of the fossil fuels that are replaced by the project (mainly kerosene for lighting) and (ii) the emissions of the biomass, if it had not been burned in the biomass generator. Without the project, this biomass would be left to decompose, producing CO_2 and CH_4 (methane). Thus, the baseline is defined to consist of fossil fuel emissions and the emissions of rotting biomass. The project's emissions consist of emissions from the burning of biomass and the remaining use of conventional fossil fuels. It is expected that the project will replace 90 per cent of the energy use; the other 10 per cent also has to be calculated as project emissions.

Table 15.2 Results of calculations of baselines and emission reductions for the Philippines case study

(all numbers in tonnes of CO_2-equivalents; numbers represent totals over entire lifetime)

Scenario:	Calculation method for baseline	Baseline emissions	Project emissions (bottom-up)[1]	Emission reduction scenario
(1) Baseline: top-down historic trend Project emissions: bottom-up	Linear extrapolation historic national trend (CO_2-eq./(GDP/cap)) and GDP/cap + rotting biomass emissions	8880	3970	4910
(2) Baseline: bottom-up historic trend Project emissions: bottom-up	Emissions of current use kerosene in village + rotting biomass emissions	7700	3970	3730
(3) Baseline: top-down projected future situation. Project emissions: bottom-up	National (GDP/cap) and (CO_2-eq./(GDP/cap)) converging to current Dutch values over the 20 years of the project + rotting biomass emissions	11210	3970 (3a) 6070 (3b)[3]	7240 (3a) 5140 (3b)[3]
(4) Baseline: bottom-up projected future situation. Project emissions: bottom-up	Emissions of current use kerosene and expected use of diesel generators village + rotting biomass emissions[2]	9260	3970	5300

Notes:

1 As noted, the project emissions were only calculated bottom-up, i.e. on a project-level basis. This estimate was provided by the project company (except for estimate (3b), which was calculated by the author). This company based the project emission calculation on its own experiences in other projects and its expertise.

2 Baseline estimate provided by the project company, based on measurements, experience from other projects and assumptions.

3 Here, it is assumed that with the project the growth rate equals the growth rate in the baseline. The number is a very rough estimate of the emission reduction. Numbers for (3a) represent the calculation as specified in case (3).

The model presented in the previous section was used to conduct four alternative baseline calculations: top-down historic trend; bottom-up historic trend; top-down projected situation and bottom-up projected situation. In Table 15.1 these were indicated as cases (1), (2), (3) and (4).[2] Figure 15.2 and Table 15.2 summarize the results, and the table includes a brief description of the method used to calculate the various baselines. Figure 15.2 shows the baseline and project emissions over the years; in Table 15.2 the totals are given for the total project lifetime.

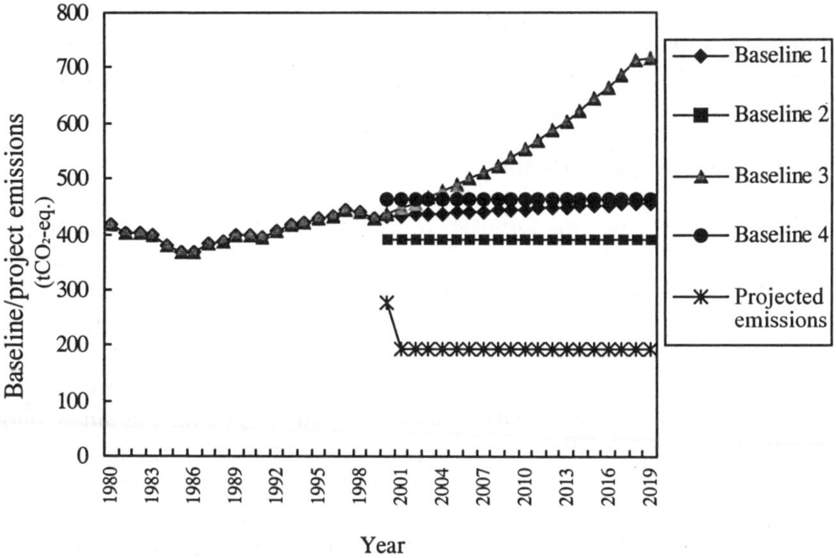

Figure 15.2 Project emissions and four baselines (tonnes of CO_2-eq.)

The baseline calculations require some extra attention. In case (1), the baseline approach is top-down and based on historic trends. It is thus based on national data for the Philippines.[3] The baseline was calculated by extrapolating both the national trend in emissions per GDP, per capita and the trend in GDP per capita. From these trends, the extrapolated historic trend of emissions per capita during the years 2000 to 2019 was derived. Finally, this latter trend was multiplied by the number of people in the village (assumed to be constant at 200) to arrive at the estimated emissions for the village over the project lifetime. The emissions from rotting biomass are then added to this number.[4] Case (3a) uses the same approach, but now the emissions per GDP, per capita and the GDP per capita are not assumed to develop according to a linear trend, but are projected to converge to the current values in the Netherlands (as a benchmark). This calculation therefore uses a strong growth *baseline* scenario. The *project* emissions, however, are based on a moderate growth scenario. It is not surprising that the resulting emission reductions are large. Assuming equal growth in the 'with project' situation gives smaller emission reductions, as represented by case (3b).

Baseline cases (2) and (4) are almost the same. Both include the emissions from the current use of fossil fuels. Lack of data meant that it was not possible to calculate a trend, so fossil fuel use is assumed to be constant over the 20 years of the project. Also, both cases (2) and (4) incorporate the emissions from rotting biomass. The difference is that case (4), reflecting a projected future situation, includes emissions from the use of inefficient diesel generators in the baseline. The project company expects these would have been used if electrification through the project had not occurred.

3.1 Emission Reductions and Costs

Table 15.2 shows that the estimates of the emission reductions of the project vary from 3730 to 5300 tonne CO_2-eq. The estimated reduction of 7240 is not considered here, since it represents an unlikely situation (that is high growth in the baseline, small growth with the project, which means the project hinders development).

Unfortunately, it was not possible to calculate the emission reduction costs per tonne under various scenarios, since no financial data for this single pilot project were made available. The larger project of which the pilot is part is expected to have a positive Net Present Value at an interest rate of 7 per cent and a moderate Rate of Return. The project is thus expected to have a negative reduction cost for greenhouse gas emission reduction. However, the expected payback period of approximately 10 years is very long.

3.2 Discussion

The baseline values that were calculated in the case study range from 7700 to 11210 tonnes CO_2-eq., thus showing a large difference. One should note, however, that the baseline and emission reduction calculations are only rough indications of the 'real' emission reductions that can be associated with the project. Many assumptions had to be made in the calculations, including a 20-year lifetime and the assumption of no leakage.

The main goal of the case study was not to assess the exact reductions, but merely to give an indication and to show the influence of varying baseline approaches. Furthermore, as was explained, the case resulting in emission reductions of 11210 tonnes CO_2-eq. is not realistic.

Despite the inaccuracies, the model approach shows that the 'real' emissions reductions of the project will probably be in the range of 5000 tonnes

CO_2-eq. Although the project has a negative reduction cost, at first sight the financial performance of the project does not seem to be very attractive.

4 CASE STUDY EGYPT: FUEL SWITCH

This case study involves a fuel switch project for a number of cities and one industrial area in Egypt. Natural gas will replace three fuel sources, namely LPG (Liquefied Petroleum Gas; for domestic and commercial use), heavy fuel oil (for industrial use) and light fuel oil (for commercial and industrial use).

The project includes two phases. In the first phase, a natural gas pipeline system is constructed from a high-pressure grid to the region under consideration. The project connects three industrial users in an industrial area and approximately 25000 domestic and commercial users to the natural gas grid. In the second phase, the industrial use of gas is expanded and an additional 58000 domestic and commercial users are connected to the grid. Both construction phases should take three years. The lifetime of the entire project is, according to the company, 20 years (including construction).

As in the Philippines case study, the model presented in section 2 was used to calculate the baseline for the project in four different ways. Again, it was inappropriate to calculate the project emissions top-down, based on national data.

Table 15.3 summarizes the results of the baseline and the emission reduction project. For the calculations we were largely dependent on information provided by the project company, for example for data on expected gas sales. The definition and calculation of baseline cases (1) and (3) is the same as in the Philippines project discussed before. Cases (2) and (4) are equal in this project. According to the model, the difference in approach between these bottom-up cases is that (2) uses the historic trend as a reference, while (4) uses a projected future situation. In the data provided by the project company, however, the projected future fossil fuel use was based on historic use, causing the two cases to coincide.

Table 15.3 shows some remarkable results: cases (1) and (3) show an increase in emissions following the project. There may be two reasons for this:

1. The project boosts energy use so much, that the emission reductions per unit of energy consumed are more than compensated by a large increase in energy consumption (large rebound effect). If this is the case, the calculated emission increase is real.
2. In calculating the baseline scenario, one or more unrealistic assumptions are used. If this is the case, the baseline scenario is unrealistic, as is the calculated emission increase.

Table 15.3 *Results of calculations of baselines and emission reductions for the Egypt case study*

Case	Calculation method for baseline	Baseline emissions	Project emissions (bottom-up)*	Emission reduction
(1) Baseline: top-down historic trend. Project emissions: bottom-up (emissions of expected gas sales)	Linear extrapolation historic national trend (CO_2-eq. per GDP per cap) and GDP per cap.	16.3	21.1	−4.8
(2)/(4) Baseline: bottom-up, projected situation is based on historic trend. Project emissions: bottom-up	Emissions of fossil fuel equivalent of anticipated gas sales; based on current use and ratio of LPG, heating oil and solar	28.5	21.1	7.4
(3) Baseline: top-down projected future situation. Project emissions: bottom-up	National (GDP per cap) and CO_2-eq. per GDP per cap. Converging to current values in the Netherlands over the 20 years of the project	18.3	21.1	−2.8

Notes:
* Project emissions were calculated as the emissions caused by the expected gas sales.
All numbers in Mtonnes of CO_2-equivalents; numbers represent totals over entire lifetime.

The first reason is unrealistic, because the expected gas sales used in the 'with project' scenario are based on current fuel use and do not incorporate a large growth in energy use. The project could only produce more emissions than the baseline situation if in the baseline emission *reductions* took place compared to the current situation. Looking at the Egyptian national trends in emission intensity of energy consumption and total emissions per capita, this, however, seems unlikely.

Then, only the two second reason remains, indicating that very likely one (or more) of the assumptions in the calculations are unrealistic. One of the assumptions indeed seems incorrect: it is implicitly assumed that the ratio of households to commercial and industrial users involved in the project is comparable to the national ratio. According to data provided, the whole region has almost 2 million inhabitants, with five major cities and *one* existing industrial zone. The project, however, covers three towns (82000 households) *and the industrial area*. Even when assuming an average household size of ten people, not even half the population in the region is involved in the project, whereas all regional industry is included. This implies that the ratio of households to industry in the project is much smaller than the national average. This leads to a serious underestimation of the baseline emissions for the project and therefore the calculation shows an emission increase caused by the fuel switch project.

It is concluded that the top-down baseline cases (1) and (3) calculated for this project are highly unrealistic. To correct this, information on regional GDP would be necessary. This information was, however, not available, so it was not possible to conduct this calculation. The model could only generate one baseline that may be considered realistic: case (2).

4.1 Emission Reduction and Costs

Since only case (2) gives a result that can be considered realistic, the emission reduction under this scenario was used in the cost effectiveness calculations. The cost effectiveness calculations are based on the first phase of the project only, since no cash flows for the second phase were available. When only phase one is executed (that is connecting three industrial, and 25000 domestic and commercial users to the natural gas grid) the resulting emission reductions are 4.70 Mtonnes CO_2-eq. over 20 years (calculated in the same way as the reductions for phases one and two in Table 15.3). For matters of confidentiality, only some general financial information is given.

Without revenues from CERs (Certified Emission Reductions, to be gained in a CDM project), the Net Present Value of the project is around 15 million US$ at a 10 per cent interest rate, with an Internal Rate of Return of over 15 per cent. The (undiscounted) payback period is slightly less than 5 years. If one assumes a CER[5] price of US$10, NPV would increase to over 30 million US$ (at 10 per cent), with the Internal Rate of Return exceeding 20 per cent. The (undiscounted) payback period would drop to 4 years.[6]

4.2 Discussion

The case study shows extremely diverging outcomes of emission reductions for top-down and bottom-up baseline scenarios. It further confirms the statement of Ellis and Bosi (1999) that the local and regional variability of the different baseline components will determine the extent to which standardization of the baseline is feasible. In the Egypt case discussed, the regional variability in industry concentration and GDP caused a standardized, top-down approach to fail. When constructing a baseline, one should be very cautious that project or regional characteristics fairly resemble national characteristics.

If the project is granted CERs, the financial performance will improve considerably, moving to a higher level of financial attractiveness. However, as the project is profitable without credits, the CDM criterion of (financial) additionality might cause problems in the certification process.

The emission reductions calculated in a top-down approach might be overstated for this project, since the top-down approach implicitly assumes that the gas sales replace all current fossil fuel use. In practice, this is unrealistic. It would be more accurate to use an estimated 'replacement percentage', if available.

5 ECUADOR CASE STUDY: WASTE MANAGEMENT

The case study for Ecuador deals with the reduction of GHG emissions of waste management in Quito, the capital of Ecuador. The study (Benitez, 2000) focuses on the various options for reducing emissions as compared to uncontrolled landfilling. Cases considered are sanitary landfilling with energy recovery and a system of integrated solid waste management. The study is based on an analysis of various options that are considered with regard to cost effectiveness and environmental impact, including emissions of GHG, liquid effluents, land degradation and recovery of resources. Valuable materials that can be obtained from the waste stream are dry recyclables, compost and energy.

Under scenario 1 of uncontrolled landfilling the private costs are low and there is no treatment of the biogas (methane) and liquid effluents. Under scenario 2 of sanitary landfilling with energy recovery the gas is collected and used for electricity generation by means of piston internal combustion and gas or steam turbines. The efficiency in conversion is in the range of 30 per cent. An alternative use of the landfill gas is to first remove the toxic contaminants of the gas and later distribute it for domestic uses (Benitez, 2000). Scenario 3, integrated solid waste treatment, includes magnetic separation of metals and manual sorting of waste on conveyor belts for paper, plastic, glass and non-ferrous metals.

Comparison of the scenarios enables us to establish the reduction of GHG emissions as compared to the baseline, which is considered to be scenario 1 and the additional net costs of reducing greenhouse gas emissions under scenarios 2 and 3.

Table 15.4 shows the additional costs and the total emission reduction for the scenarios 2 and 3 as compared to scenario 1.

Results show that both a sanitary landfill and an integrated solid waste treatment system can reduce GHG emission at a cost of about US$25 per tonne. At the same time, both systems result in additional positive externalities, such as reducing liquid effluents and the use of space. Trading of GHG by means of CDM offers scope for increased cost effectiveness if emission reduction options in industrialized countries cost more than US$25 per tonne CO_2 eq.

To what extent the baseline in this project is acceptable depends on the judgement whether it is likely that in the coming 20 years a system of sanitary landfill or integrated solid waste treatment would be introduced in Quito, even if no climate policies played a role. This judgement is necessary to define a proper baseline on a project basis. It is, however, extremely difficult – and the case study clearly emphasizes this aspect! – to identify a scientific approach in order to establish whether or not these systems would be introduced without climate policies.

Table 15.4 Total costs and emissions reduction for scenarios 2 and 3 compared to baseline scenario 1

Scenario	Total costs of emission reduction (million US$)	Total emission reduction (million tonne of CO_2 eq.)	Cost of CO_2 emission reduction (US$/tonne CO_2 eq.)
Sanitary landfill	206	8	25.8
Integrated solid waste treatment	316	12.9	24.5

Source: Benitez (2000).

6 MEXICO CASE STUDY: AGROFORESTRY

The case study on forestry and agroforestry measures in the Central Highlands of Chiapas in Mexico is based on de Jong et al. (2000). In their research they analyse the costs of reducing greenhouse gas emissions by forestation, fallow management and agricultural and pasture management in Mexico. The study takes into account the private costs of these management options and the opportunity costs of agricultural income for gone.

The study pays extensive attention to the risk of carbon leakage, that is the option that additional deforestation will take place at another location, outside the project boundaries. If this type of carbon leakage occurs, a project-based approach should pay due attention to it, in order to avoid a severe overestimation of the emission reductions of the project.

The project shows that reforestation can result in relatively large reductions of CO_2 emissions at relatively low costs of US$5–20 per tonne. Within this cost range, forestry and agroforestry measures in the relevant study area could mitigate from 1 to 42 tonnes of carbon, with a maximum economic supply of carbon sequestration of around 55 tonnes of Carbon at US$40 per tonne of Carbon. If sufficient warranty could be granted that indeed the forestation project is additional and that no carbon leakage would take place, the project would be useful for CDM. Also in this case, however, it is not very clear whether reforestation would occur anyway or not, if no climate policies were implemented. Again, it is emphasized that it is difficult - if not impossible - to identify an unbiased scientific method to make a judgement about the additionality of the project.

7 DISCUSSION AND CONCLUSIONS

In this chapter we have discussed basic approaches for establishing the baseline for GHG emissions for JI and CDM, together with four case studies. In many cases the project approach was used. The various case studies show the following.

1. Top-down approaches may lead to completely different results compared to bottom-up approaches. The different baseline approaches taken in the case studies lead to different results for the same projects. The Philippines project shows that calculating different baseline scenarios can give a fair indication of the emission reductions of a project, although the various approaches can lead to a difference in calculated emission reductions of up to 30 per cent.[7] In the Egypt project, the different scenarios lead to extreme differences in outcomes. As was discussed, this is caused by a di-

vergence in project and national characteristics, making it impossible to set a top-down, standardized baseline. Although there is a clear reason for the diverging outcomes, they cannot be solved without expanding the data requirements of the model.

2. Carbon leakage and the issue of lifetimes (related to additionality) cause problems in both types of approach.

The case studies show that it is basically impossible to establish a well-defined baseline on the basis of a scientific approach: even after a careful baseline study, it is not clear when and where new technologies or management practices will be introduced, for what reasons and how much carbon leakage will occur. This is why we call for a political choice regarding the preferred method of establishing baselines, or at least regarding minimum requirements or standards. Such a decision will not solve the problems mentioned above, but will at least provide clarity on the margin of error that can be expected in baseline calculations.

When deciding on the baseline methodology or approach, decision makers should keep the consequences of their choice in mind. The approach used to determine the baseline for emissions has consequences for the project's transaction cost, transparency and administrative feasibility (including data, monitoring and reporting requirements), as well as for its environmental additionality (and the emission reductions claimed). These are indicated as 'performance criteria' (Ellis and Bosi, 1999). Besides these, the approach chosen also influences other factors, presented schematically in Figure 15.3. Each baseline approach has its own advantages and disadvantages. Basically, the question to consider is, which approach will score best on the performance criteria. The difficulty is that in practice there is a trade-off between the different criteria, which implies that values or weights will have to be attached to the criteria before a final choice can be made. Ideally, baselines should be credible, transparent, simple and inexpensive to set up (Ellis and Bosi, 1999).

Specification of how a baseline should be established has other advantages. First, there is always a heavy reliance on data provided by the investor in calculating emission reductions. Although the top-down baselines themselves hardly rely on data that are provided by the investor, the calculations of project emissions do. Thus, even if the baseline itself does not rely on investor data, calculation of the project emissions does, allowing for strategic behavior by the investor. Specification of the baseline method will reduce cheating. Second, since various baseline approaches lead to different outcomes for emission reductions, the costs per unit of emission reduction will also be affected by the approach taken. It is thus difficult to assess the cost effectiveness and efficiency of the JI and CDM instruments as long as

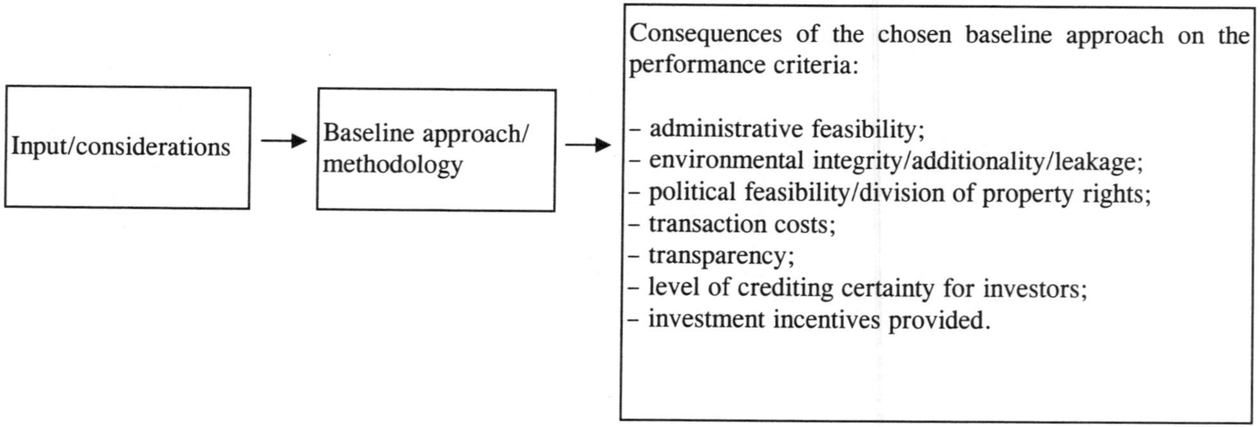

Sources: OECD (1999), Ellis and Bosi (1999), Meyers (1999) and Bandsma et al. (2000).

Figure 15.3 Consequences of baseline approach chosen

there is no standardized approach to assess the emission reductions of the project and the reduction costs. Without such a standardized approach, international cost effectiveness cannot be achieved.

We recognize that, given the various stages of development in various regions of the world and given the introduction of new technologies at different times in the future, it will be extremely complicated to identify well-defined and non-arbitrary baselines. The complicating factor in this is that the choice will influence the division of 'property rights' between countries and therefore should be of great concern to all parties, not just Annex I countries. It is therefore absolutely essential that all countries participate in the international negotiations on climate change.

Our results emphasize the complications in establishing baselines and the additionality of GHG emission reduction projects. Top-down baseline setting requires a large number of very arbitrary assumptions about future economic and technological development and can only be applied if the key characteristics of the project resemble national ones. Therefore, we strongly recommend that baselines be established on a project basis, while paying due attention to the project boundaries and the risk of carbon leakage. The decision on how to set the baseline will have a direct impact on the actual levels of emissions reduction, because the set of rules will determine how much emission leakage may occur. Strategic behaviour of stakeholders (both host and investors) may call for top-down baseline calculations, to check the results. We recommend that the acceptance of various projects be tightly controlled and that all parties be treated fairly and in a consistent manner. If this is not done, the discussions on the baseline and the eligibility of a project in one country and the refusal in another country will have very negative impacts on the willingness to participate in future greenhouse gas emissions reductions treaties, both for industrialized and developing countries.

NOTES

1. Activities Implemented Jointly. This concerns projects carried out under the pilot phase for both Joint Implementation and the Clean Development Mechanism. This pilot phase ended by 2000. Under AIJ, emission reductions could not be used to meet national commitments.
2. The project emissions were calculated in a bottom-up manner only, since it was considered inappropriate and (from lack of data) impossible to calculate the project emissions in a top-down manner.
3. Derived from the internet sites of the US Energy Information Agency; www.eia.doe.gov
4. This figure was provided by the project company and is derived from emission factors found in literature.
5. 1 CER is 1 ton of CO_2-eq. The ultimate price of emission credits depends on supply and demand. Several simulation studies have been conducted to gain insight in the level of the market price for credits once the market is up and running. For example, ECN (1999) estimates a price of \$4–\$15 (energy options only), and in Van der Gaast and Woerdman (1999) several studies are reviewed with price estimates of \$10 (G-cubed model, 1998, in

Van der Gaast and Woerdman, 1999), $23 (EIA, 1999, in Van der Gaast and Woerdman, 1999) and $40 (CEA, 1998, in Van der Gaast and Woerdman, 1999). Furthermore, the World Bank uses a price of US$10 in its Prototype Carbon Fund (World Bank, 2000).

6. Other assumptions: the CERs are paid for at the end of the year in which they are realized. The exchange rate used is US$0.34 per £E (Egyptian Pound). Exchange rate at 16 May, 2000.
7. Excluding the unrealistic (outlier) scenario (7a).

REFERENCES

Bandsma, J., W. Van der Gaast and C.J. Jepma (2000), 'Baseline: Criteria and Issues; Overview of discussion and results of previous workshops on the criteria of baselines', Prepared for the Expert Workshop to Develop Initial Guidelines on Baseline Determination, Amsterdam, 17–19 January, Paterswolde (The Netherlands), Joint Implementation Network.

Benitez, P. (2000), 'Economic and Environmental Aspects of Municipal Solid Waste Management in Quito, Ecuador', Wageningen University, Environmental Economics Group, Wageningen, The Netherlands.

ECN (Netherlands Energy Research Foundation ECN) (1999), 'Potential and cost of Clean Development Mechanism options in the energy sector; Inventory of options in non-Annex I countries to reduce GHG emissions', in co-operation with AED (Alternative Energy Development Inc, Silver Spring, USA) and SEI (Stockholm Environmental Institute, Boston, USA), on behalf of the Netherlands Development Co-operation (DGIS), Petten, ECN, December.

EIA (Energy Information Agency) (2000), 'Various statistics' at www.eia.doe.gov and related internet pages, United States Government.

Ellis, J. (1999), 'Experience with Emission Baselines under the AIJ Pilot Phase', OECD/IEA Information Paper, Paris: OECD.

Ellis, J. and M. Bosi (1999), 'Options for project emission baselines', OECD/IEA information paper, Paris: Organisation for Economic Co-operation and Development and International Energy Agency.

Gaast, W. van der and E. Woerdman (1999), 'A review of the cost-effectiveness of AIJ: projecting JI and CDM credit prices', draft report, the Netherlands: University of Groningen/Joint Implementation Network.

Jepma, C.J., W.P. Van der Gaast and E. Woerdman (1998), 'The Compatibility of flexible instruments under the Kyoto Protocol', Carried out under the Dutch National Research Program on Global Air Pollution and Climate Change, the Netherlands: Joint Implementation Network.

JIRC (Joint Implementation Registration Centre) (2000), *Setting a Standard for JI and CDM; recommendations on baselines and certification, based on AIJ experience,* The Hague: JIRC.

Jong, B.H.J. de, R. Tipper and G. Montoya-Gómez (2000), 'An Economic Analysis of the Potential for Carbon Sequestration by Forests; Evidence from Southern Mexico', *Ecological Economics*, **33** (2), pp. 313–27.

Leeuw, G.J. de (2000), 'The Clean Development Mechanism for Dutch Investors; A preliminary study of options, conditions and issues', thesis, University of Twente /Cartesius Institute/Wageningen University, Environmental Economics Group.

Meyers, S. (1999), 'Additionality of Emission Reductions from Clean Development Mechanism Projects: Issues and Options for Project-level Assessment', Berkeley, CA: Ernest Orlando Lawrence Berkeley National Laboratory.

Michaelowa, A. (1998), 'Joint Implementation - the Baseline Issue; Economic and Political Aspects', *Global Environmental Change*, **8** (1), pp. 81–92.

OECD (Organisation for Economic Co-operation and Development) (1999), 'Status of Research on Project Baselines under the UNFCCC and the Kyoto Protocol', OECD working papers, **VII** (25), Paris: OECD.

Stewart, R. (Lead Author) (1999), *The Clean Development Mechanism: Building International Public–Private Partnership; A preliminary examination of technical, financial and institutional issues* (based on the deliberations of the Ad Hoc Working Group on the CDM), New York: United Nations.

Woerdman, E. (1997), 'Internationale Onderhandelingen over Klimaatbeleid: Emissiereductie, Joint Implementation en Verhandelbare Emissierechten', *Milieu; tijdschrift voor milieukunde*, **5**, Meppel, Boom Uitgeverij.

World Bank (2000), Web-site Prototype Carbon Fund, available online at www.prototypecarbonfund.org/html/vcc.htm

16. Economic issues related to design of a domestic permit trading system[1]

Carolyn Fischer

1 INTRODUCTION

In Kyoto in 1997, the US government agreed that between 2008 and 2012 it would reduce average annual emissions of greenhouse gases (GHGs) to 7 per cent below 1990 levels. Participants in the climate policy debate are expressing considerable and growing interest in the use of an 'emissions trading' policy to achieve these limits. By using tradable emissions permits, as opposed to command-and-control regulation, society benefits as a whole from having environmental goals achieved at lower cost.[2] Over time, GHG controls implemented by permit trading will induce investments in the development and diffusion of new technologies (such as renewables and enhanced energy efficiency) for reducing GHG emissions. Several recent experiments with emissions trading to control other pollutants, notably the phasing out of lead in gasoline and the ongoing reduction of SO_2 emissions from electric utilities under Title IV of the US Clean Air Act, suggest that an emissions trading policy is very promising.[3] However, there is still considerable debate and some confusion about how a GHG emissions trading program would be organized and operated in practice.

In this chapter, we address a number of questions that must be considered in designing and implementing a domestic GHG trading system in the US. These include who and what is covered by regulation (what sources and gases); the commodity to be traded and the trading system; how rights to emission permits are defined; how the system is to be initiated and changed over time; how to deal with different kinds of uncertainties; and the interactions of permit trading with the tax system. While some of these questions have been discussed in previous studies, this chapter draws these issues together and adds new insights concerning uncertainty, intertemporal consistency, market institutions and interactions with the tax system. Our focus is exclusively on domestic policy; we do not consider how emissions trading or other GHG control policies should function in an international context.

Throughout we consider the overall cost effectiveness of alternative approaches (including administrative costs), their distributional implications and political economy considerations.

Our fundamental conclusions are:

1. The basic system should be a 'cap-and-trade' program with the following properties:
   - An 'upstream' point of enforcement;
   - Bankable permits;
   - Allocation by government auction, with the revenue used to offset existing tax distortions and to provide carefully targeted assistance to those most adversely affected.
2. If permits are not entirely allocated by auction, methods must be careful to minimize adverse effects:
   - The rules for altering permit allocations should create the proper incentives for the private sector to respond to risk while also limiting opportunistic government behavior.
   - If permits are issued gratis, they should be taxed on their market value.
   - If permit liabilities can be offset by emission abatement activities conducted abroad, transfer prices at market permit rates should be required to prevent tax shifting.

2 DESIGN PRINCIPLES FOR A GHG TRADING SYSTEM[4]

2.1 What does a Permit Cover?

Direct regulation of all or even most GHG emissions is all but impossible in practice: looking only at CO_2 emissions from fossil fuel use, the most important GHG, there are literally scores of millions of sources. In addition, highly precise methods for directly measuring CO_2 emissions are economically practical only for large boilers that can be equipped with continuous emission monitors. One alternative is to control fossil fuel inputs to the economy. Fortunately, the CO_2 emitted from fossil fuel is essentially perfectly correlated with the carbon content of the fuel. Therefore, a reasonably accurate measure of emissions can be obtained by keeping track of quantities of fossil fuel used together with their carbon content.

A 'cap-and-trade' program involves issuing a quantity of homogeneous permits (that is, not sector- or GHG-specific) equal to the total level of emissions the Protocol allows. Individuals are free to buy and sell permits subject to whatever bookkeeping requirements are needed for assuring compliance. The 'commodity' traded is the opportunity to emit a unit of

GHG (measured in terms of carbon content released) once, rather than defining the commodity as a stream of GHG emissions over time. This approach, which is used already to regulate US utilities' SO_2 emissions, provides maximum flexibility and liquidity in the permit market. 'Transaction costs' – the costs of identifying trading partners and effectuating trades – can be very low and 'derivative' transactions such as forward, futures and options contracts can develop. Other human-induced greenhouse gas emissions (such as methane from coal mines and landfills, or HFCs now used in lieu of CFCs in air conditioners) could be added to the cap-and-trade system (based on calculations of their relative contributions to global warming), depending on the capacity to measure or infer emission balances. Sources, sinks and gases not included directly, because of difficulties in tracking aggregate balances, could be incorporated in ancillary project-specific efforts to create credits that could be 'imported' into the core program.[5]

2.2 When can a Permit be Used?

Each permit can be used after a given date. For example, permits for the first budget period would be dated 2008, those for the second period, 2013. Permits could be used or banked for future use. Banking is a key feature of an efficient trading system, since it allows for efficient arbitrage of marginal costs when these are rising over time as emission targets tighten or energy demand grows. Regulators could also issue or sell permits ahead of their 'use after' dates, and emitters and fuel suppliers could engage in forward contracting to assemble portfolios of permits. In particular, it may be useful to initiate provision or sale of permits for the initial commitment period (2008–2012) somewhat beforehand, so as to allow price information to develop and to build confidence in the trading and monitoring institutions themselves. Experience with the US SO_2 program suggests that confidence is enhanced by establishing in advance the key provisions: how trading will work and how results will be monitored.

An important related question is how any pre-2008 reductions of GHG emissions or fossil fuel use would 'count' in terms of compliance with subsequent regulatory limits. Any such crediting depends on the establishment of credible, internationally recognized baselines for assessing the early reductions. Parry and Toman (2000) discuss this issue in depth.

2.3 Who is Required to use Permits?

Where to enforce the use of carbon-content permits is primarily a question of administrative convenience. Upstream enforcement, focusing on the point of entry of carbon to the domestic market, would involve the lowest number of entities to regulate. For example, it would be possible to control natural gas entering interstate pipelines, crude oil (domestic and imported) entering refineries and coal sales from mines or processing plants.[6] Such a system would require the supervision of fewer than 2000 actors.[7]

One might also consider various hybrids combining regulation of downstream CO_2 emissions from certain users of carbon (such as large boilers, like in the US SO_2 program) with upstream regulation of other fossil fuel flows. This approach requires more complex record keeping and enforcement and it has greater risks of sources 'leaking' from the system. More importantly, given reasonably competitive product and factor markets (including progress toward a competitive restructuring of the US electric utility industry), a downstream approach will have the same ultimate effects on fossil fuel and other prices as an upstream program. This corollary of the basic proposition – that the ultimate incidence of a tax is independent of where it is applied – is frequently misunderstood by proponents of downstream-based GHG emissions trading. The main substantive argument for a hybrid approach instead appears to be the ability to allocate emission permits to various energy users; we return to this point below.

2.4 How would Permits be Enforced?

Once a capacity for measuring or reliably estimating individual actors' CO_2 emissions or fossil fuel sales is established (a feasible but not trivial task), it is necessary only to keep track of covered sources' permits relative to fuel flows or emissions. There would be no need to keep track of specific transactions in the trading registry, other than the creation of project-specific non-CO_2 emission credits imported into the core cap-and-trade program. Any shortfall between required and actual permits would be deterred by stiff financial penalties and a requirement to cover the shortfall. Permits would be fully fungible within the five-year commitment period, but each year firms would be required to show they hold permits equal to or greater than their emissions. This annual reporting would avoid the risk of large carbon debts when firms go bankrupt and could provide useful information to the permit market.

3 OPTIONS FOR ALLOCATING PERMITS[8]

Before any trading program could commence, the permits would first have to be allocated. The choice of mechanism can have a great impact not only on the distribution of the permit rents, but also on administrative ease, transparency and the overall efficiency of the system. We consider three main methods: auctions, gratis allocations ('grandfathering') and updating allocations ('output-based').

3.1 Auctions

The government could simply require prospective permit holders (and anyone else who wished to) to bid for permits in an auction. An active secondary market could and (unless foolishly prohibited by government) would evolve that would allow sources to adjust their permit holdings to changes in circumstances between auctions. Auctions could be held periodically (for example, quarterly). The government could auction more permits than are expected to be used within the period, to allow 'borrowing'.

A great deal of research and a reasonable amount of practical experience is available for designing an efficient auction for permits.[9] This institution would guarantee access to permits for small agents. Importantly, an auction would raise considerable revenue for the government.[10] This revenue could be used in several ways:

- To mitigate the adverse effects of GHG control for certain groups of businesses or workers,
- To help finance innovation in technologies for GHG reduction (such as new energy sources or carbon storage technologies), for adaptation to climate change, or
- To benefit taxpayers as a whole through reductions in other taxes.

As we discuss further below, this last approach has the potential to reduce the net cost of GHG control substantially, assuming the tax cuts are broad-based and efficiency-increasing.

3.2 Gratis Distribution

The opposite alternative to an auction is some form of gratis allocation. Gratis allocation provides a tool for distributing the scarcity rents created by limits on fossil fuel use or GHG emissions. It reduces the resistance of those who would be in an auction having to pay for all their carbon or emission permits. However, the beneficiaries from a political decision to allocate permits to certain parties may not correspond to those who ultimately face the most adverse economic effects from imposition of the program.

One often-discussed strategy for gratis allocation is a form of 'grandfathering', whereby regulated emission sources or fuel suppliers are given permits in proportion to their historical emissions or fossil fuel sales. This 'lump-sum' method redistributes the rents from emissions to the recipients, typically the shareholders of the favored firms. However, an unconditional giveaway of emission permits has the same effects on marginal costs and ultimately the same price and behavioral effects as with a permit auction.[11] For example, employees that are unable to change jobs easily in response to lower labor demand in specific sectors such as coal, or households that cannot easily reduce energy demand in response to higher energy prices will suffer such effects under any system. Gratis allocation also deprives the economy of the opportunity for reducing existing distortionary taxes. Furthermore, it may reduce somewhat the incentive for technical innovation: a cost-saving innovation with widespread applications will tend to drive down the permit price, reducing the value of grandfathered permit holdings.[12]

3.3 Output-Based Allocation

Gratis allocation could occur in a number of other ways, including allotments conditioned on behavior. In these cases, the price and behavior effects may not be the same as with auctioned permits. One approach is output-based allocation, as proposed by Lashof (1997). Each sector would be granted a fixed number of permits and within each sector, firms would receive permits proportional to their share of their industry's output. This method ensures that future players will receive free permits, as the allocations are automatically updated over time. However, distribution of permits in this manner amounts to an inefficient subsidy on output. This subsidy mitigates part of the rise in output prices, providing less of a signal to reduce consumption of the products associated with emissions. Firms are thus encouraged to invest more in further reducing the carbon intensity of output, rather than reducing output itself. To achieve the same emissions cap, given the output subsidy, permit prices would have to be higher, as would overall compliance costs.[13]

Implementing such a program would also prove more complicated and could exacerbate inefficiencies. Information about market shares, not just carbon use, would have to be gathered. As market shares are determined industry by industry, setting different overall allocations for each industry creates different subsidy rates for each, resulting in more inefficient shifting of production and abatement activity between industries.

3.4 Recommendation

For these reasons of efficiency, we strongly prefer auctioning.[14] However, we do recognize that an important force behind the drive for gratis allocation is the creation of political support for environmental policy. To accomplish this goal, it would be possible to start with a mixture of gratis allocation and auctions and gradually phase out the gratis allocation. Bovenberg and Goulder (2000) address the likely mix of permit giveaways and auctions for a politically feasible CO_2 permit program, finding that the fraction of rents necessary to compensate major industries for lost profits is relatively small. Windfall losses apply only to investments that are 'sunk' and irreversible; as capital naturally turns over, future investments will be redirected, taking the emissions policy into account. Phasing into an auction process ensures that the distribution of future rents will subsequently pass through the familiar filter of the fiscal policy process, rather than a separate forum of interests seeking emission rents.

The McKibbon–Wilcoxen proposal for emission endowments also recognizes the need for a constituency supporting the environmental policy. They recommend grandfathering an indefinite stream of permits in part to fossil fuel industries and in part to individuals. Individuals play many roles in the economy – as consumers, shareholders, permit asset holders and taxpayers – and they will be concerned with environmental policy, whether the accompanying emission rents are distributed to one role or another. By allocating endowments as a perpetual asset, owners have a vested interest in the stringency of future caps, not just concerned with the current rents and costs of environmental policy. However, that vesting comes at a cost to future generations, who will find it more difficult and expensive to redistribute those rents. In the next section, we discuss other trade-offs with respect to the duration of endowments.

With respect to what to allocate, in light of the argument in the previous section, one should distribute upstream fossil fuel carbon permits, *not* emission permits. Any gratis allocation of these permits to downstream actors would then be bought back by the upstream actors who directly face regulation (for example, refineries), making the flows of rents transparent. However, it can be argued that a more direct, transparent and flexible way to achieve redistribution of the emission rents is through the standard fiscal policy process – through transfers of revenues from a permit auction to whatever groups of shareholders, consumers, workers or taxpayers are deemed worthy of receiving them.[15]

Implicit in this analysis is an assumption that the right to allocate, and thereby the initial property rights to the emissions, is vested with the government. By its power to regulate, the government is creating private property out of a previously common one. Clearly, governments have the legal power to levy taxes and distribute the revenues and the effects of an emissions permit policy can be achieved equally by an emission tax policy. However, this economic equivalence may not extend to legal equivalence. If initial rights to the commons are not vested with the government but with existing users (such as the individual citizens who breathe the air, or the carbon users who pollute it), re-allocation and restriction of those property rights by the government could become difficult. In the United States, the acid rain program successfully restricted SO_2 emissions, created a market for them and designated grandfathered permit allocations; however, the same legislation declared that the permits were not 'property' in order to avoid 'takings' issues should the cap need to be changed. On the other hand, while the SO_2 program may seem to create tempered optimism, stronger precedent does exist for the federal government creating and auctioning rights to use ethereal common property: the broadcast spectrum has been allocated in just this manner.

4 INTERTEMPORAL FLEXIBILITY AND UNCERTAINTY

4.1 Banking and Borrowing

As noted previously, banking of unused permits is important for improved efficiency when regulatory targets are such that marginal control cost is rising over time faster than the relevant rate of interest. By the same token, long-term borrowing against future allocations to meet nearer-term targets will be attractive if the price of permits otherwise is expected to rise more slowly than the rate of return on other assets, indicating a disproportionate short-term burden of controlling GHG emissions relative to the longer term. This could occur because the composition of energy-using capital is more flexible in the longer term than in the shorter term and because tougher short-term requirements provide relatively less opportunity to embed technical improvements over time.

Modeling studies suggest that emitters can stabilize atmospheric concentrations of GHGs at far lower cost (well over 50 per cent savings), with little or no adverse climate change implications in the short-to-medium term, if they can maintain significantly higher emissions in the next couple of decades than are allowed under the Kyoto Protocol. During these decades, they would begin to install higher-efficiency capital that will allow much lower emissions subsequently.[16] These models provide an economic argument for permit borrowing (or long-term allocation of permits without restriction on advance use). Critics of these analyses have argued that postponement of short-term emissions reductions, by lowering energy prices below what would have occurred with tighter standards, will also retard some induced innovation and learning-by-doing that is assumed by proponents to be available in the future.[17] Reaping the cost savings pointed to by advocates of borrowing thus might require increased government commitments to R&D.[18]

Critics of borrowing have also raised more fundamental questions related to time consistency. What is to stop emitters from rapidly depleting their endowments of permits in anticipation that current regulators will not be willing to impose sanctions? If future regulatory actions to enforce repayment of borrowed permits are not seen as credible, then needed investments in R&D will not be undertaken. In addition, what would prevent future regulators, who will themselves have incentives to defer control costs, from not going along with the higher control cost burdens that a current decision to defer abatement would place on them? This problem might be countervailed in two ways: (1) Efforts to discover and diffuse, perhaps with government support, significantly lower-cost ways to limit GHG emissions would make postponing abatement less attractive. (2) Selling permits well in advance (but allowing use in advance) would give private holders an incentive to resist subsequent efforts by the government to add permits to the system and create capital losses on banked holdings.

Because the potential cost savings from providing increased intertemporal flexibility are so great, it is important to continue to investigate ways that such flexibility might be credibly implemented in practice. The government could also attempt to approximate the results of intertemporal borrowing by announcing a sequence of tighter emission limits over time. However, there is still a credibility problem related to whether the government will follow through on a commitment to tighten emissions limits and this uncertainty will deter R&D. The McKibbon–Wilcoxen proposal offers an example of options for dealing with these issues: by distributing the entire stream of permits in the initial endowments, those asset holders will form a constituency in favor of tightening the cap. Those permits cannot be banked or borrowed, however. Instead, marginal abatement costs are equalized over time, at least in the short term, by fixing permit prices for 10-year periods.

4.2 Bearing Risks

A GHG trading program must also deal with various uncertainties over time. 'External' factors reflect changes in costs or compliance requirements outside the control of the domestic government. Aggregate national GHG emissions ceilings could be revised due to such international influences as changes over time in the perceptions of the risks of climate change and the ongoing process of negotiation of global emissions targets and national obligations under the Framework Convention. For example, a discovery that climate change is more threatening than it seems today could cause national GHG budgets to be reduced to phase down emissions more quickly. Given any emission target, the cost of compliance faces inherent uncertainties in technological progress, business cycles and many factors affecting energy demand.

'Internal' uncertainties reflect the government's own political or revenue-based motivations for changing the policy. The credibility (or lack thereof) of advance policy commitments is one major source of uncertainty, as alluded to in the previous section. Other sources arise even if the Kyoto targets are met. If permit allocations are not credibly fixed, the potential for reallocation of gratis quota among sectors is a source of internal uncertainty. For example, downstream sectoral quotas might be reduced when a sector has shown particular capacity to reduce costs.

These uncertainties call for responses that are to some extent at cross-purposes. Putting the external risk on the private sector creates uncertainties in their investment planning. However, these risks are unavoidable and emitters, not the government, are in the best position to bear them through a portfolio of investments in existing and new technologies that provide flexibility if quotas fall (and options to emit more if they should happen to rise).[19] On the other hand, internal risks unnecessarily increase investment uncertainty and they can undercut the incentives for permit market participants to undertake desirable investments in lowering the cost of emissions control.

By their nature, hybrid upstream/downstream programs with gratis allocation seem more prone to internal risks than an upstream auctioned approach, since differentiation of regulation or allocation by sector will lead to competing pressures for preferential treatment. A permit policy needs to spell out carefully the circumstances under which emissions allocations can be altered, in order to limit opportunistic regulatory behavior, while recognizing the sovereignty of the government over GHG emissions and its need to respond to ongoing international discussions on GHG reductions.

Kopp et al. (1998) propose a modification to the Kyoto Protocol to limit external uncertainties. They would combine a cap-and-trade system for GHG emissions with a maximum price for permits, at which the government promises to sell unlimited permits. Thus, if compliance costs are low, the benefits of reduced emissions are assured and if costs prove to be high, the safety valve takes effect and assures costs do not rise to extreme levels. Furthermore, by using the revenue generated by safety valve sales to pay for GHG reduction in developing countries, some of the forgone benefits of emissions reduction are recaptured.

5 PERMIT TRADING AND INCOME TAXATION

Any permit system has to operate within the context of the existing domestic and foreign tax systems. Each system will have an effect on the efficiency of the other, raising important design issues for GHG trading. For domestic income tax policy, of particular concern is the treatment of banked permits over time: general inefficiencies in the taxation of capital income raise some specific issues for intertemporal emissions abatement choices. For multinational companies, the choice of how to implement international offsets can affect their ability to avoid corporate income taxes by shifting the location of their abatement activities. Finally, with respect to overall efficiency of the tax systems, the choice for generation and use of permit revenues can affect the burden created by other taxes, particularly labor income taxes.

5.1 Deductions and Capital Gains

Permits are potential assets as well as inputs to production. This dual nature poses a challenge in choosing proper tax treatments and achieving efficiency in the permit market. When market conditions favor banking of permits (that is, when abatement costs are expected to rise sufficiently in the future to warrant holding permits for later use), the taxation of capital gains can introduce distortions in permit markets. One distortion is a 'lock-in' effect that stimulates excessive banking of permits. Given the choice of (a) selling a permit and paying tax on capital gains accrued to date, then investing the sale proceeds and (b) holding the permit in expectation of further capital gains, which will only be taxed when the permit finally is sold, existing permit holders will tend to defer permit sales that might otherwise be economically efficient.

The tax system also creates a kind of reverse lock-in: allowance purchase costs cannot be deducted as inputs until use, which deters purchase of allowances in advance. However, because of the first lock-in effect, the deferral of deductibility is crucial in order to deter inefficient tax avoidance. Without this rule, taxes could be saved by purchasing and immediately deducting allowance costs and deducting interest in funds borrowed to finance the purchase, while only paying tax on the accrued return when the asset is sold. Still, the same kind of tax arbitrage can occur when permit buyers' purchase and fully deduct higher-cost permits while holding permits with a lower cost basis for accrual of gains. This particular scenario can be avoided by a type of FIFO (first-in, first-out) rule.

Inefficiencies arise when permits face different effective tax rates across different individuals. For example, an owner of a permit would require a lower rate of return to holding permits to offset the lock-in effect and sell today rather than hold the asset. However, with different cost bases, different owners will require different premia to overcome the particular lock-in they face and sell today. On the other hand, for a buyer to be willing to purchase and hold a permit, she would require a higher rate of return, since she cannot deduct the cost until later. No permit price path can simultaneously make all these actors indifferent to buying, selling and holding.

Gratis allocation creates more of a bias toward excessive permit banking, the first lock-in effect. This will tend to inefficiently 'tilt' the permit price path downward: permits today will be more expensive and the greater volume of banked permits will lower future price expectations. The excessive tilting of the price path in turn will shift abatement activities inefficiently toward the present. On the other hand, with auctioned permits the bias is against holding, since the deduction must be deferred. In this case, for banking to become worthwhile, the price path must inefficiently tilt up to make up for the tax distortion; correspondingly, some abatement will be inefficiently deferred.

Ultimately, reducing these distortions would require broader reform of capital gains taxation.[20] Simply eliminating the taxation of capital gains on permits would not eliminate the problems; as long as permits are deductible, gains must be taxed. Otherwise, two permit users could swap permits, with each deducting the full value and neither being taxed on the proceeds. Exempting permits from taxation completely would treat firms symmetrically and get rid of this incentive to 'churn'. However, differences in income tax brackets among individuals and corporations mean that different investors still will face different tradeoffs between permits and other assets and distortions will remain. If permits were tax-exempt then people with high marginal tax rates would invest in them, driving down the rate of return to the low after-tax rate of return these individuals would reap on other assets; consequently, the lower rate of return would cause excessive amounts of early abatement effort.

The important question is regarding the magnitude of the interactions between the emissions trading program and these features of the tax system – how substantial are the distortions to abatement decisions, and how large might be the potential loss of tax revenues through arbitrage activity? Recent evidence in the literature on capital gains taxation suggests that the decision relating to when people choose to realize capital gains is not that sensitive to tax rates over the long term, implying a relatively small lock-in effect.[21] However, gratis permits pose a larger potential lock-in effect than most assets: since the cost basis for gratis permits is zero, they in effect instantaneously accrue the market value. The incentive to defer gains by holding onto permits is not only the postponement of taxes on the price increases, but also the postponement of taxes on the initial permit price itself. Short of auctioning permits, this problem could be mitigated by taxing the recipient on the market value of the allocation, thereby allowing the initial basis to be the market rate and not zero. Firms would then deduct that market value for the permits they use, so at tax time they would effectively only be liable for gains tax on the gratis permits they did not use.

5.2 Multinationals and Corporate Income Taxation

A large body of literature indicates that corporate income taxation does have significant influence on a wide range of activities by multinational corporations, including foreign direct investment, borrowing, transfer pricing, dividend and royalty payments, R&D activity, exports and location choices.[22] Location and extent of abatement activity will be added to this list if foreign abatement efforts can be used to offset domestic emission liabilities. The issue is to encourage cost-effective foreign abatement and discourage non-productive efforts to reduce tax liabilities.

Multinational companies operating in countries with higher corporate income tax rates than their country of origin can reduce their tax burdens and increase net profits by realizing more gains at home and incurring costs abroad. Opportunities to offset domestic emissions liabilities with foreign abatement activities can also create opportunities for shifting profits and lowering taxes. For example, corporate income taxes are significantly higher in Japan than in the United States. A US multinational with operations in Japan, if allowed to offset some domestic permit requirement with abatement performed in Japan, would save taxes by incurring abatement costs in Japan and reaping permit savings at home.[23] If this form of tax arbitrage was permitted, the multinational would shift more abatement activity to Japan than would be justified by the actual cost savings.

The best way to prevent these incentives is to make the transactions as transparent as possible. Appropriate transfer pricing makes profit shifting difficult. Consequently, offsets should be required to be valued at the domestic (or international) market permit price. Being required to pay the going rate for the offsets, the parent would incur the appropriate costs and the subsidiary the appropriate benefits. Abatement then occurs abroad only if it is economically justified and truly profitable.[24] Having a well-functioning domestic permit market can thus also foster appropriate incentives for emissions reductions abroad. An international permit market with clear prices could further enhance the transparency necessary for tax enforcement and efficient global emissions reduction.

5.3 Interaction with Tax Distortions

In addition to specific interactions with the tax code, a permit system can interact with the tax system through the generation of auction revenue. Since leisure goes untaxed, any tax on consumption goods or income lowers the effective real wage and distorts the labor supply decision. Consequently, any environmental policy that raises the cost of carbon-intensive goods exacerbates the pre-existing distortions in the labor market. This type of tax interaction is best mitigated by collecting permit rents through auctions and using the revenues to lower distorting labor taxes.[25] Coupling an auctioned permit system with broad-based income tax reform not only minimizes additional labor market inefficiencies but also lessens tax distortions to the permit system itself.

6 CONCLUDING REMARKS

These conclusions that the permit system should be upstream, auctioned and that auction revenue should be recycled are still controversial. However, if one assumes a longer-term national commitment to potentially costly GHG emissions control there is both a fair amount of time for regulatory institutions to evolve and gain public acceptance and a pressing need to choose the best institutions possible. Without such a commitment, which is currently lacking in the political arena, any discussion of regulatory options is somewhat beside the point. Only time will tell if our proposal gains currency in the seven years between now and the start of binding commitments under the Kyoto Protocol. Many important practical questions need to be addressed before a GHG trading system in the United States can be constructed and implemented. However, the deployment of a relatively efficient and transparent system seems within our grasp, especially once key issues related to who benefits from the revenues created by such a system are resolved. A well-developed domestic trading system in turn would provide a good foundation for taking advantage of the potential opportunities afforded by GHG trading on an international scale.

NOTES

1. For distribution at the 2000 Taipei Conference on Policies for Greenhouse Gases Reduction and Pollution Control in Asian Pacific. This chapter is adapted from a series of papers by Carolyn Fischer, Suzi Kerr and Michael A. Toman, 'Using Emissions Trading to Regulate US Greenhouse Gas Emissions', RFF Climate Issues Briefs nos 10 and 11, Washington, DC: Resources for the Future, June 1998.
2. The relative advantages of emissions trading as a quantity-based instrument and carbon taxes as a direct price-based instrument is an important question (Pizer, 1997). In this chapter, we limit our attention to quantity-based trading policies.
3. Hahn and Hester (1989), Burtraw (1996), Kerr and Maré (1997), Stavins (1998).
4. A substantial literature is emerging on this issue (Smith et al., 1992; Hahn and Stavins, 1995; Swisher and Ellsworth, 1997; Swift et al., 1997; Hargrave 1998; Festa 1998). A related literature has also emerged in connection with international GHG policies (Tietenberg and Victor, 1994; Fisher et al., 1996; Stewart et al., 1996).
5. In addition, some arguments have been made for including producers of energy-using capital equipment in a trading system (for example, by requiring vehicle manufacturers to hold permits equal to the expected lifetime emissions of their vehicles). However, this approach controls only the performance of new equipment, has no effect on the utilization of equipment and creates a bias that encourages uneconomic life extension of older, less efficient equipment. It can also lead to double regulation of emissions from energy-using capital equipment. Over the long term, price signals provided by a fuels-based program will effectively guide both equipment purchase and utilization decisions. It is certainly possible that market failures associated with energy-using equipment (information unavailability, institutional barriers to transmission of effective price signals and the like; see for example Jaffe and Stavins, 1994) need to be addressed to improve the efficiency of capital purchase decisions. However, these problems are best attacked directly where they occur.

6. A variety of adjustments would need to be made. The basic system would not capture the relatively small amount of energy used in producing and initially transporting the fossil fuels to the regulatory control points. A rough adjustment might be made for that energy by requiring permits in excess of the actual carbon content of the fuel. It would be necessary also to adjust for non-combustion uses of fossil fuel inputs (such as chemical feedstocks, taking into account that some of the carbon in these materials eventually may escape to the environment as wastes are burned). Biomass energy supplies could be included by requiring biomass supplies to be permitted for their carbon content but then providing credits for the carbon sequestered in the growing of the feedstock. Similar crediting provisions would be needed for other deliberate carbon-sequestration activities, notably reforestation.
7. Cramton and Kerr (1998), Hargrave (1998).
8. We use the term 'carbon permits' here because it is common parlance, but a broader and more accurate term would be 'GHG permits' to reflect the existence of GHGs besides CO_2.
9. See Cramton and Kerr (1998).
10. To illustrate, US emissions of carbon from fossil fuel burning in 1990 were calculated to be 1374 million tonnes. Reducing this by 7 per cent by 2008 would imply emissions of about 1280 tons. (The US is obliged only to reduce all GHGs by an average of 7 per cent by 2008–2012.) If the price of a GHG permit were $50/tonne of carbon emission, then 1280 million tonnes of permits in the US would be worth $64 billion per annum.
11. Grandfathering is sometimes criticized as being biased against new sources, which must pay for their emissions permits while existing sources obtain theirs for free. However, unless there are deeper competition problems, existing firms will have neither the motivation nor the capacity to discriminate against new entrants. For their part, entrants can and should come into an industry if and only if they are more efficient than incumbents, taking into account the opportunity cost of GHG permits. It is possible that small new sources could be disadvantaged because of imperfect capital markets that limit their access to finance. But this problem is best rectified by addressing the sources of any capital market distortions
12. See Fischer, Parry and Pizer (1998).
13. See Fischer (1999) for more discussion.
14. See also Cramton and Kerr (1998) for further discussion.
15. To avoid unnecessary inefficiency, these payments would have to be structured so as not to become output or consumption subsidies.
16. See Richels and Edmonds (1995); Manne and Richels (1996, 1997).
17. See Grubb (1997) and Ha-Duong et al. (1997).
18. A more complex question, which we do not address here, is what level of public and private R&D actually is desirable and the extent to which a desirable level will be undertaken when a policy like emissions trading is imposed.
19. Banked allowances should be sacrosanct; without this assurance, a valuable means for improving the efficiency of emissions control over time will be lost.
20. Auerbach (1992).
21. See Gravelle (1991), Burman and Randolph (1994) and Zodrow (1995).
22. See Hines (1996) for an overview.
23. This situation occurs when the US multinational has 'excess tax credits'; that is, its foreign tax payments are higher than its domestic tax liability for its foreign income.
24. The analysis is more complicated if both countries have separate permit programs and offsetting is limited, but the basic results hold. See Fischer (2000).
25. See Goulder (1995), Parry (1995), Parry et al. (1997).

REFERENCES

Auerbach, A.J. (1992), 'On the Design and Reform of Capital Gains Taxation', *American Economic Review*, **82** (2), pp. 263–67.

Bovenberg, A.L. and L.H. Goulder (2000), 'Confronting Distributional (and Political) Impacts of CO_2 Abatement Policies: What Does It Cost?', RFF Discussion Paper 00-27, Resources for the Future, Washington, DC, July.

Burman, L.E. and W.C. Randolph (1994), 'Measuring Permanent Responses to Capital Gains Tax Changes in Panel Data', *American Economic Review*, **84**, pp. 794–809.

Burtraw, D. (1996), 'The SO_2 Emissions Trading System: Cost Savings Without Allowance Trades', *Contemporary Economic Policy*, **14** (1), pp. 79–94.

Cramton, P. and S. Kerr (1998), 'Tradable Carbon Permit Auctions: How and Why to Auction not Grandfather', Discussion Paper 98-34, Resources for the Future, Washington, DC, June.

Festa, D. (1998), 'US Carbon Emissions Trading System: Some Options that Include Downstream Sources', draft manuscript, Center for Clean Air Policy, Washington, DC.

Fischer, C. (1999), 'Rebating Environmental Policy Revenues: Output-Based Allocations and Tradable Performance Standards', RFF Working Paper.

Fischer, C. (2000), 'Multinational Taxation and International Emissions Trading', RFF Working Paper.

Fischer, C., I.W.H. Parry and W.A. Pizer (1998), 'Instrument Choice for Environmental Protection when Technological Innovation is Endogenous', RFF Discussion Paper 99-04, Washington, DC: Resources for the Future, October, forthcoming in *Journal of Environmental Economics and Management*.

Fisher, B.S., S. Barrett, P. Bohm, M. Kuroda, J.K.E. Mubazi, A. Shah and R.N. Stavins (1996), 'An Economic Assessment of Policy Instruments for Combating Climate Change', in J.P. Bruce, H. Lee and E.F. Haites (eds), *Climate Change 1995: Economic and Social Dimensions of Climate Change*, Contribution of Working Group III to the Second Assessment Report of the Intergovernmental Panel on Climate Change, Cambridge: Cambridge University Press, pp. 397–439.

Goulder, L.H. (1995), 'Effects of Carbon Taxes in an Economy with Prior Tax Distortions: An Intertemporal General Equilibrium Analysis', *Journal of Environmental Economics and Management*, **29**, pp. 271–97.

Gravelle, J. (1991), 'Limits to Capital Gains Feedback Effects', *Tax Notes*, 22 April.

Grubb, M. (1997), 'Technologies, Energy Systems and the Timing of CO_2 Emissions Abatement: An Overview of Economic Issues', *Energy Policy*, **25** (2), pp. 159–72.

Ha-Duong, M., M.J. Grubb and J.C. Hourcade (1997), 'Influence of Socioeconomic Inertia and Uncertainty on Optimal CO_2-Emission Abatement', *Nature*, **390**, pp. 270–73.

Hahn, R. and G. Hester (1989), 'Marketable Permits: Lessons for Theory and Practice', *Ecology Law Quarterly*, **16**, pp. 361–406.

Hahn, R. and R. Stavins (1995), 'Trading in Greenhouse Permits: A Critical Examination of Design and Implementation Issues', in H. Lee (ed.), *Shaping National Responses to Climate Change*, Washington, DC: Island Press.

Hargrave, T. (1998), 'US Carbon Emissions Trading: Description of an Upstream System', unpublished manuscript, Washington, DC: Center for Clean Air Policy.

Hines, J. (1996), 'Tax Policies and the Activities of Multinational Corporations', NBER Working Paper No. 5589.

Jaffe, A.B. and R.N. Stavins (1994), 'The Energy Paradox and the Diffusion of Conservation Technology', *Resource and Energy Economics*, **16** (2), pp. 91–122.

Kerr, S. and D. Maré (1997), 'Transaction Costs and Tradable Permit Markets: The United States Lead Phasedown', draft manuscript, University of Maryland, College Park, Md.

Kopp, R., R. Morgenstern, W. Pizer and M. Toman (1998), 'Cheap Emissions Reductions: Use 'Em or Lose 'Em', Washington, DC: Resources for the Future, 23 February, at http:\\www.weathervane.rff.org/features/feature023.html

Lashof, D. (1997), 'Output-Based Allowance Allocation', Manuscript, Natural Resources Defense Council, September.

Manne, A.S. and R. Richels (1996), 'The Berlin Mandate: The Cost of Meeting Post-2000 Targets and Timetables', *Energy Policy*, **24** (3), pp. 205–10.

Manne, A.S. and R. Richels (1997), 'On Stabilizing CO_2 Concentrations – Cost-Effective Emission Reduction Strategies', draft manuscript (April), Stanford University, Stanford, CA.

Parry, Ian W.H. (1995), 'Pollution Taxes and Revenue Recycling', *Journal of Environmental Economics and Management*, **29**, pp. S64–S77.

Parry, Ian W.H. and M. Toman (2000), 'Early Emissions Reduction Programs: An Application to CO_2 Policy', Discussion Paper 00-26, Washington, DC: Resources for the Future, June.

Parry, Ian W.H., R.C. Williams III and L.H. Goulder (1997), 'When Can Carbon Abatement Policies Increase Welfare? The Fundamental Role of Distorted Factor Markets', Discussion Paper 97-18, Washington, DC: Resources for the Future.

Pizer, W.A. (1997), 'Prices vs. Quantities Revisited: The Case of Climate Change', Discussion Paper 98-02, Washington, DC: Resources for the Future, October.

Richels, R. and J. Edmonds (1995), 'The Economics of Stabilizing Atmospheric CO_2 Concentrations', *Energy Policy*, **23** (4/5), pp. 373–78.

Smith, A.E., A.R. Gjerde, L.I. DeLain and R.R. Zhang (1992), 'CO_2 Trading Issues, Vol. 2: Choosing the Market Level for Trading', report to the Environmental Protection Agency, Washington, DC: Decision Focus Inc.

Stabins, Robert N. (1998), 'What Can We Learn from the Grand Policy Experiment? Positive and Normative Lessons from SO_2 Allowance Trading', *Journal of Economic Perspectives*, **12** (3), pp. 69–88.

Stavins, Robert (1998), 'What Can We Learn from the Grand Policy Experiment? Positive and Normative Lessons from SO_2 Allowance Trading', *Journal of Economic Perspectives*, **12** (3), pp. 69–88.

Stewart, R., J. Wiener and P. Sands (1996), 'Legal Issues Presented by a Pilot International Greenhouse Gas Trading System', November, Geneva: United Nations Conference on Trade and Development.

Swift, B., Jay Pendergrass, Eric Wilkinson, Vanessa Reeves and Carol Riegelman (1997), 'Implementing an Emissions Cap and Allowance Trading System for Greenhouse Gases: Lessons from the Acid Rain Program', Washington, DC: Environmental Law Institute, September.

Swisher, J. and Amy Ellsworth (1997), 'Analysis of the Potential for a Greenhouse Gas Trading System for North America', unpublished report, Boulder, CO: Econergy International Corporation.

Tietenberg, T. and D. Victor (1994), 'Possible Administrative Structures and Procedures', in *Combatting Global Warming*, Geneva: United Nations Conference on Trade and Development.

Zodrow, G.R. (1995), 'Economic Issues in the Taxation of Capital Gains', *Canadian Public Policy*, **21** (Supplement), pp. 27–57.

Index